Effective Stress and Equilibrium Equation for Soil Mechanics

Effective Stress and Equilibrium Equation for Soil Mechanics

Longtan Shao and Xiaoxia Guo

Department of Engineering Mechanics,
Dalian University of Technology, Dalian,
Liaoning, China

Shiyi Liu

School of Resources and Civil Engineering,
Northeastern University, Shenyang,
Liaoning, China

Guofeng Zheng

Department of Engineering Mechanics,
Dalian University of Technology, Dalian,
Liaoning, China

CRC Press
Taylor & Francis Group
Boca Raton London New York Leiden

CRC Press is an imprint of the
Taylor & Francis Group, an **informa** business
A BALKEMA BOOK

Published by: CRC Press/Balkema
Schipholweg 107c, 2316 XC Leiden, The Netherlands
e-mail: Pub.NL@taylorandfrancis.com
www.crcpress.com – www.taylorandfrancis.com

First issued in paperback 2020

CRC Press/Balkema is an imprint of the Taylor & Francis Group, an informa business

ISBN 13: 978-0-367-57245-7 (pbk)
ISBN-13: 978-1-138-09231-0 (hbk)

Library of Congress Cataloging-in-Publication Data

Applied for

Typeset by Apex CoVantage, LLC

Contents

Preface

Terzaghi (1925, 1936) proposed the concept of effective stress and the effective stress principle, which established the basis of soil mechanics. However, Terzaghi neither provided an accurate definition of effective stress nor explained its physical meaning. He only indicated "*All the measurable effects of a change of the stress, such as compression, distortion and a change of the bearing resistance, are exclusively due to changes in the effective stresses.*" Therefore, researchers deduced the effective stress according to the test results of shear strength and volumetric change and proposed many types of modification formulae for effective stress equation. There were some other researchers who believed effective stress is interparticle stress or soil skeleton stress. However, following their definition of interparticle stress or soil skeleton stress, they could not obtain Terzaghi's effective stress equation accurately. On the other hand, Terzaghi's effective stress equation is only applicable for saturated soils, not for unsaturated soils. Bishop (1959) expanded this equation to unsaturated soils and proposed the effective stress equation for both saturated and unsaturated soils. Similarly, since the definition and physical meaning of effective stress were not clear, confusion was encountered in the explanation of effective stress and determination of the coefficients in effective stress equation.

This book aims to indicate the physical meaning of effective stress and illustrate that effective stress is the soil skeleton stress, excluding the effect of pore fluid pressure. The effective stress equation characterizes the equilibrium relationship of the inner forces in soil mass and owns the uniform expression for saturated and unsaturated soils, which is permanently established. It indicates the interparticle stress and soil skeleton stress, excluding pore fluid pressure, are consistent with the effective stress and also elucidates why effective stress governs the strength and deformation of soils.

To illustrate the physical meaning of effective stress and derive the effective stress equation, we applied the methods of continuous medium mechanics to conduct internal force analysis on the free bodies of the soil skeleton, pore water and pore air separately and deduced the equilibrium differential equation for each phase of the soil. When the free body of the soil skeleton is taken to conduct force equilibrium analysis, the stress on the surface of the free body has two parts: one part is induced by pore fluid pressure, which only includes normal stress, and the other part is due to all the other external forces (excluding pore fluid pressure), which includes normal stress and shear stress. Comparing the equilibrium differential equation of the soil skeleton to that of total stress, the relationship among total stress, pore fluid pressure and soil skeleton stress due to the external forces (excluding pore fluid pressure) can be obtained. It can be found that this relationship under full-saturated condition is Terzaghi's effective stress equation. Consequently, we call the soil skeleton stress due to all the external forces, excluding pore fluid pressure, effective stress; illustrate the reason why the effective stress equation exists theoretically and on what condition the effective stress determines the deformation and strength of a soil; and simultaneously indicate the effective stress equation is valid, no matter whether for saturated soils or for unsaturated soils.

In this way we can conclude that Terzaghi's effective stress equation does not need to be modified and that the effect of pore fluid pressure has to be excluded from internal force analysis when using soil skeleton stress or interparticle stress to explain effective stress. The perspective that effective stress should be deduced according to the test results of strength and volumetric change is inappropriate. For unsaturated soils, we can determine the coefficients in Bishop's effective stress expression and give them explicit physical meanings.

On the other hand, the expression of Darcy's law is obtained in this book based on the equilibrium differential equation of pore water, which, in laminar seepage, is the condition that the moving resistance is linearly proportional to the velocity of pore water. For unsaturated soils, the expression of the relationship among coefficient of permeability, degree of saturation and water potential is derived.

The key point of this book is the interpretation of effective stress, based on the equilibrium differential equation of component phases. By carefully reading the book, it can be found that the equilibrium differential equation, rather than effective stress, has fundamental and foundational meaning for establishing the theoretical system of soil mechanics. The equilibrium differential equation is essential for soil mechanics, which is the core content in this book. The effective stress equation is a natural result of the equilibrium differential equation, and the equation of Darcy's law and the coefficient of permeability of unsaturated soils are only specific applications of the equilibrium differential equation of pore fluid. The issues discussed in the last chapter of this book are related to the equilibrium differential equation or effective stress. These issues are important to soil mechanics, while most of them are controversial. The discussion on these issues aims to eliminate these controversies and help the readers better understand the role of equilibrium equation and master the concept of effective stress. For simplicity, this book only provides the equilibrium differential equation of soil mechanics in static condition, which are readily extended to dynamic condition.

Actually, it is very simple to derive the equilibrium differential equation by taking the soil skeleton as the study subject and considering the effect of pore fluid pressure and external forces separately to conduct internal force analysis. However, no one has done this before. Biot (1941) obtained the consolidation equation by introducing the effective stress equation into the total stress equilibrium equation, which actually reflects the equilibrium of the soil skeleton. Zienkiewicz and Shiomi (1984) and the other researchers also derived the equilibrium equation of the soil skeleton via inducing the effective stress equation into the total stress equilibrium equation in the stress-strain computation of soils. Fredlund and Morgenstern (1977) took the soil skeleton, pore water, pore air and contractile skin (surface tension skin) as analysis subjects independently and deduced the equilibrium equation and the stress state variable that controls the equilibrium of soil structure, whereas he did not directly derive the equilibrium differential equation of the soil skeleton. In his research, the soil structure included the soil skeleton and contractile skin; the equilibrium differential equation of the soil skeleton was obtained by subtracting the equilibrium equation of pore water, pore air and contractile skin from that of total stress. In the theory of mixtures, the soil skeleton, pore water and pore air were treated as study subjects separately to derive their individual equilibrium differential equation. However, in the internal force analysis of the soil skeleton, it did not distinguish the two different equilibrium force systems of pore fluid pressure and external forces. In its definition of soil skeleton stress, the effects of the external load and pore fluid pressure are both included, and the physical meaning of effective stress is consequently hidden.

Most of the content in this book is the reorganization and summary of previously published research work by the authors. Before this, Shao (1996) proposed the interaction principle for multiphase media, including (1) in the force and deformation analysis of soil mass, as a multiphases medium, each phase of the soil skeleton, pore water and pore air should be treated as an independent analysis subject; (2) the interaction between interphase forces can be described by a pair of force and counterforce in the force analysis of each phase medium and (3) the deformation and motion of the soil skeleton, pore water and pore air are only determined by their own state variables, boundary condition and initial condition. He derived Terzaghi's effective stress equation for saturated soils and the effective stress equation for unsaturated soils (Shao, 2000, 2011a, 2011b, 2012; Shao and Guo, 2014; Shao *et al.*, 2014) based on the interaction principle for multiphase media. He also discussed some issues (Shao *et al.*, 2014), such as whether the effective stress is real stress of soil mass, whether the effective stress equation needs to be modified, the definition and formula of seepage force, the stress state variable and effective stress equation of unsaturated soils, and whether the contractile skin of unsaturated soils needs to be treated as an independent phase.

This book discusses the definition of the "soil skeleton" and clarifies that the soil skeleton should include a portion of bonding water in pores. Why a portion of bonding water should be treated as a component of the soil skeleton is discussed in Section 1.4. There is a coefficient, degree of saturation, before the term of matric suction in the effective stress equation for unsaturated soils derived from the equilibrium differential equation. However, the stress expression in the formula of the shear strength of unsaturated soils obtained by Vanapalli and Fredlund (Vanapalli *et al.*, 1996) via experiments has the same form as the effective stress equation we derived, only the corresponding coefficient is effective degree of saturation (refer to Section 3.2.2 in this book). This implied that the parameter for the degree of saturation in the effective stress equation should be effective degree of saturation. This might not be a new perspective, which, however, has neither been clarified in previous theoretical research of soil mechanics nor been practically applied. Apparently, this perspective is crucial to the theoretical research of soil mechanics, at least may change the definition of dry soils, and may affect the stress-strain characteristic and the research on the constitutive relationship of fine particle soils. It is not an easy-to-complete task to prove that the pore water corresponding to residual water content should be treated as a component of the soil skeleton. The micro-scale research on the structure of soils appears very important herein. Unfortunately, the current research results are inadequate to verify this perspective (Section 1.4 in this book), while the available references about this aspect are limited. The corresponding research can be conducted in micro-scale and macro-scale. The micro-scale research includes finer and more accurate tests on soil-water structure and interaction, molecular dynamic simulation and the research of soil-water combined structural model. The macro-scale research can be preceded with the characteristics of strength and deformation change of fine particle soils.

During the derivation of the equilibrium equation via taking the free body of the soil skeleton to conduct the internal force analysis, pore fluid pressure and the effect of the other external forces are separately considered for two reasons. First, when we take the free body of the soil skeleton that does not include pore fluid to conduct internal force analysis, the forces acting on the soil skeleton by pore fluid should appear, which includes the forces due to equilibrated fluid pressure and non-equilibrated fluid pressure. The force due to equilibrated fluid pressure is self-equilibrated with fluid pressure. It can and should be considered independently; and the force due to non-equilibrated fluid pressure should be represented with a couple of action and reaction forces, i.e., seepage forces. Second, different from the effect of the other external forces, the uniform pore fluid pressure only induces volumetric change of soil particles and contributes to the shear strength of soils on the contact surfaces between particles (as shown in Section 3.2 in this book).

The definition and physical meaning of effective stress are found by separately considering the effects of pore fluid pressure and the other external forces. This finding is helpful in eliminating the always-existing controversy on effective stress and unifying the research on saturated soil mechanics and unsaturated soil mechanics.

Compared to all known theories of soil mechanics, this book presents some other new perspectives and results. Except for the definition and physical meaning of effective stress, these new results also include (1) the development of Terzaghi's effective stress principle (Section 3.4); (2) the effective stress equation of unsaturated soils (including nearly saturated soils containing absorbed air bubbles) (Section 3.1); (3) the correction of the formula of seepage force (Section 4.3); (4) the preliminary verification of the relationship between effective stress and shear strength and deformation of unsaturated soils (Section 3.3); (5) the theoretical formula of the permeability coefficient of unsaturated soils; and (6) the escape condition for occluded air bubbles in water (Section 4.4). As of yet, the test verification for the effective stress of unsaturated soils has not been finished, and the verification for the theoretical formula of the permeability coefficient of unsaturated soils and the escape condition for occluded air bubbles in water is under way.

The research in this book assumes that the soil is homogeneous, ignoring the shear strength of pore water (expect for the pore water that is treated as a component of the soil skeleton) and the interaction force between pore water and pore air induced by the relative flow. It should also be noted that the work reported here has been established based on the following important postulations: (1) the strength of soil mass is that of the soil skeleton, and (2) the deformation of soil mass is that of the soil skeleton. These two postulations are of great significance for the research of soil mechanics.

Acknowledgements

During the writing and preparation of this book, we wish to express our gratitude to the people who have contributed to this book. It was the "Alfried Krupp und Bohlen und Halbach Stiftung", who provided the financial support for the first author to stay in University of Karlsruhe for conducting the research on effective stress of unsaturated soils as a visiting professor under the guide of Professor G. Gudehus. The roots of the ideas expressed in this book lie in the work of the first author during his staying in University of Karlsruhe in 1998–1999. In addition, we express our sincere thanks to Professor Sai K. Vanapalli and D. G. Fredlund for giving the stress expression in the formula of shear strength of unsaturated soils obtained via experiments, which reminds us that the pore water corresponding to residual moisture content should be considered as a part of soil skeleton. We also thank Mr. Yazhou Zou, a senior engineer at Universität der Bundeswehr München, Professor Zigang Zhang of the Institute of Physics of Chinese Academy of Sciences, Professor Guangxin Li of Tsinghua University, Professor Changfu Wei of Guilin University of Technology, Professor Liangtong Zhan of Zhejiang University, Professor Ning Lu of Colorado School of Mines, and the graduates students of the Soil Mechanics Laboratory, Department of Engineering Mechanics, Dalian University of Technology for their contributions and discussions. In particular, we thank Dr. Gang Liu of Dalian University of Technology, for her deriving the formula of seepage force according to Figure 4–7 and her assistance in preparing the figures about force analysis on free body. Moreover, we express our sincere thanks to Dr. Mo Zhang, a lecturer at Dalian Jiaotong University and Dr. Changqing Cheng, an assistant professor at State University of New York, respectively, for their English translation and correction.

We acknowledge many scholars for giving us data from their published works as follows:

Khalili, N., Geiser, F. and Blight, G. E. (2004). Effective stress in unsaturated soils: review with new evidence, International Journal of Geomechanics, (4): 115–126.

Alonso, E. E., Pereira, J. M., Vanunat, J. and Olovella, S. (2010). A microstructurally based effective stress for unsaturated soils, Géotechnique, 60(12): 913–925.

Zhan, L. T. and Ng, C. W. (2006). Shear strength characteristics of an unsaturated expansive clay, Canadian Geotechnical Journal, 43: 751–763.

Miao, L. C., Cui, Y. J. and Cui, Y. (2015). Hydromechanical behaviors of unsaturated soils, Journal of materials in civil engineering, 27(7): 1–9.

Hossain, M. A. and Yin, J. H. (2010). Shear strength and dilative characteristics of an unsaturated compacted completely decomposed granite soil, Canadian Geotechnical Journal, 47: 1112–1126.

Many thanks are also due to the following organization and the researcher for permissions for use the indicated figure: Figure 5.39, Advanced Soil Mechanics, Braja M. Das, CRC Press, 2008.

The financial support received from the National Natural Science Foundation of the Republic of China through grant 51479023 is gratefully acknowledged.

Finally, we thank CRC Press for the publication of this book.

Chapter 1

Introduction

While soil has played a vital role since the early stage of the development of human society, the science of soil mechanics has a rather short history. The earliest documented theoretical research on soil mechanics can be traced back to the 1770s. In 1773, Coulomb researched material strength and developed the Coulomb's law of shear strength, based upon which Mohr established the Mohr-Coulomb strength theory of soil. This theory has since become the groundwork for the strength and failure study of soil and provided theoretical framework for computation of soil compression, bearing capacity of foundation soil and stability analysis on soil slopes. In a report published in 1776, Coulomb discussed soil pressure theory based on the equilibrium analysis of soil wedge and developed the computation methods for soil pressure of retaining walls. Rankine investigated soil pressure theory using plastic equilibrium analysis of the soil body in 1857. Darcy proposed Darcy's seepage law through extensive laboratory experiments in 1856 and laid a solid foundation for the percolation theory and seepage flow mechanics. Boussinesq and Flamant proposed the displacement and stress distribution theory of isotropic and homogeneous semi-infinite surfaces under vertical concentrated loading and linear loading in 1885 and 1892, respectively. In the early 20th century, according to the plastic equilibrium principle, Prandtl investigated the process of hard objects pressed into relatively soft, isotropic and homogeneous materials and derived the ultimate bearing capacity equation for soil. Later on, Terzaghi, Meyerhof, Vesic and Hansen further improved this equation and proposed their own versions of ultimate bearing capacity equations for foundations.

Despite the numerous outcomes and achievements on the study of soil mechanics, those studies were scattered and not well-organized in a systematic way before Terzaghi, who is considered a paramount figure in establishing the subject of soil mechanics, mainly in two respects. First, soil was classified into clay and sand according to physical properties, and the various laws, principles and theories on soil mechanics were summarized to form a basic framework for the study of soil mechanics. Second, effective stress principle for saturated soils and the one-dimensional consolidation theory were established. *Soil Mechanics*, published by Terzaghi in 1925, is the earliest work systematically discussing the knowledge system of soil mechanics. He published *Theoretical Soil Mechanics* in 1943 and *Soil Mechanics in Engineering Practices* with Peck in 1948 and primarily built a relatively complete knowledge system of soil mechanics and geotechnical engineering. Therefore, Terzaghi is considered the founder for the subject of soil mechanics. Based on Terzaghi's effective stress principle and consolidation theory, Biot developed the basic equilibrium equation in soil mechanics, i.e., Biot's consolidation equation. The effective stress principle along with Biot's consolidation equation has become the cornerstone for soil mechanics.

1.1 Effective stresses

1.1.1 *Effective stresses of saturated soils*

Terzaghi first proposed the concept of effective stress in 1936, with the original quote:

> The stresses in any point of a section through a mass of earth can be computed from the total principal stresses n'_{I}, n'_{II} and n'_{III} which act in this point. If the voids of the earth are filled with water under a stress n_w, the total principal stresses consist of two parts. One part, n_w, acts in the water and in the solid in every direction with equal intensity. It is called the neutral stress. The balance, $n_{\mathrm{I}} = n'_{\mathrm{I}} - n_w$, $n_{\mathrm{II}} = n'_{\mathrm{II}} - n_w$ and $n_{\mathrm{III}} = n'_{\mathrm{III}} - n_w$, represents an excess over the neutral stress n_w and it has its seat exclusively in the solid phase of the earth.
>
> This fraction of the total principal stresses is called the effective principal stresses or equal values of the total principal stresses, with the effective stresses dependent on the value of n_w. In order to determine the effect of a change of n_w at a constant value of the effective stresses, numerous tests were made on sand, clay and concrete, in which n_w was varied between zero and several hundred atmosphere pressure. All these tests led to the following conclusions, valid for the materials mentioned.
>
> A change of the neutral stress n_w produces practically no volumetric change and has practically no influence on the stress conditions for failure. Each of the porous materials mentioned was found to react a change of n_w as if it were incompressible and as if its internal friction were equal to zero. All the measurable effects of a change of the stress, such as compression, distortion and a change of the bearing resistance, are exclusively due to changes in the effective stresses, n_{I}, n_{II} and n_{III}. Hence the investigation of the stability of a saturated body of earth requires the knowledge of both the total and the neutral stresses.

This statement by Terzaghi revealed the following facts: (1) traditionally, the stress of soils people calculated is the total stress; (2) the total stress of soils is composed of effective stress and pore water pressure; and (3) the volume and strength change induced by pore water pressure are too small to be observed, whereas the effective stress is the cause for the deformation and strength change of soils.

The validity of the effective stress principle for saturated soils has been adequately verified by the work of many researchers. Bishop (1959) stated that the development of the effective stress principle to a position of importance to the civil engineer depended on three requirements: (1) laboratory methods to measure the relevant soil properties accurately and easily; (2) theories to relate the laboratory measurements to the conditions met with in the actual full-scale problems, and suitable design methods based on these theories; (3) field measurements to compare the predicted behavior with actual performance. From the point of view of engineering practice, the effective stress principle satisfied the three requirements, for all types of soils. The two simplest implications of the effective stress principle are: (1) that volumetric change and deformation in soils depend not on the total stress applied but on the difference between the total stress and the pressure set up in the fluid in the pore space. This leads to the expression $\sigma' = \sigma - u$; (2) that shear strength depends not on the total normal stress on the plane considered but on the effective stress. This may be expressed by the equation $\tau_f = c' + \sigma' \tan \varphi'$. Bishop further considered the effect of the particle-particle contact area and found that the effective stress is not completely equal to the difference between the total stress and pore water pressure, but determined by the contact area between particles.

Afterwards, the soil mechanics researchers Jennings and Burland (1962) stated Terzaghi's effective stress principle in the form of two aspects. First, changes in volume and shear strength of a soil are due exclusively to changes in effective stress, which they verified with a series of consolidation tests and full compression tests of nearly saturated soils. Second, the effective stress in a soil is defined as the excess of the total applied stress over the pore pressure. They also stated that this principle is of great significance and undoubtedly applicable to saturated soils.

Before Terzaghi, the stress of the soil body was generally considered as the total stress, i.e., the stress of the whole body of soils (the mixture of the soil skeleton and pore fluid). Although it is straightforward to compute the stress of soils and consistent with the thinking habits of researchers to use the soil body as the analysis subject, it is difficult to reflect the physical nature of the strength and deformation of the soil body. The solid skeleton of soils forms the structure of the soil body, and thus the strength and deformation of the soil body are actually the strength and deformation of the soil skeleton. However, the total stress is not the stress of the soil skeleton, and does not determine the strength and deformation of the soil body. There is no intrinsic corresponding relationship between the total stress and the strength and deformation of the soil body. The proposition and establishment of the concept of effective stress made it possible for us to master the essence of the strength and deformation of soils, set up the correlation between the strength and deformation of soils and the stress, build up the stress-strain constitutive relationship of soils, and further construct and complete the knowledge system of soil mechanics.

The effective stress principle is considered the most significant principle in soil mechanics. Jennings and Burland (1962) stated,

> The validity of this principle for fully saturated soils is now well established and has proved invaluable in the prediction of the behavior of such soils. Up till now the validity of the first proposition, upon which the main practical significance of the effective stress principle relies, does not appear to have been questioned.

Bishop and Blight (1963) claimed that the effective stress principle provides a satisfactory knowledge base not only for understanding the characteristics of strength and deformation but also for practical engineering designs. This has been broadly accepted. Mitchell (1993) considered the effective stress principle of saturated soils by Terzaghi as the "key-stone" for soil mechanics, where "key-stone" is the stone on the top of an arch structure. The importance of this principle can be illustrated by this metaphor. Li (2011) stated that

> The subjects of Terzaghi's one-dimensional seepage-consolidation theory in classic soil mechanics, the Biot consolidation theory, the drained and undrained strength in soils, the pore water pressure coefficient by Skempton, the computation of the gravity stress and additional stress of underwater soils, deformations of soils caused by seepage, the pressure of water in soils (i.e., uplift pressure and lateral pressure), the pre-compressed seepage-consolidation of bases, the stability analysis of slopes in hydrous condition by limit equilibrium method were all established based on the effective stress principle. Terzaghi's effective stress principle is the symbolic theory for that the soil mechanics can become an independent mechanical subject.

Although the effective stress principle is very important and widely accepted, its adoption in soil mechanics is not without flaw. First, the feasibility and rationality of the effective stress principle for saturated soils have been repeatedly proved by laboratory experiments and geotechnical engineering practices. However, the well-accepted expression of effective stress for unsaturated soils still remained unavailable until now. Second, the concept of effective stress was proposed by Terzaghi based on observation and empirical experience without any theoretical basis, and thus the definition is not clear. It was only stated that effective stress is equal to the total stress, subtracting pore water pressure, and the deformation and strength of soils are determined by effective stress. However, it was not explained why these two statements hold. As a result, the effective stress was a virtual physical quantity for most researchers to date. At the same time, the concept and principle of effective stress have always been discussed and disputed since they were proposed.

1.1.2 Main academic perspectives and some comments on effective stress

Skempton (1961) interpreted effective stress via analyzing the forces between soil skeleton particles. Effective stress was interpreted based on the deformation-equivalence or the shear strength-equivalence according to the fact that an infinitesimal amount of the soil body causes isometric volumetric change or the equivalent principle of shear strength. The rationality of the expressions in these two equivalent conditions was also verified through tests. He testified which expression was relevant to the volumetric change of soils due to the total stress and pore water pressure and examined which expression satisfied the Mohr-Coulomb equation of drained shear strength.

The research by Skempton (1961) later inspired many other researchers. They kept the same perspective that the effective stress should be the volumetric change- or shear strength-equivalent stress of the soil body. Based on this viewpoint, researchers proposed different expressions of effective stress and modified equations. However, the disputation has never ended.

Bishop and Blight (1963) further reformatted Skempton's equivalent stress equations of volumetric change-equivalence and shear strength-equivalence with one uniform formula, i.e., $\sigma' = \sigma - ku_w$, where k is a parameter corresponding to the volumetric change-equivalence and shear strength-equivalence, respectively. According to this definition, the effective stress is a function of the total stress and pore water pressure, which governs the stress change during the change of volume and shear strength. Bishop also thought,

> Terzaghi's effective stress principle is a special case of the equation $\sigma' = \sigma - ku_w$, when k is equal to 1. Although it is a great approximation from the engineering point of view, it does not have generality, as a special case.

Nur and Byerlee (1971) also considered the compressibility of soil particles and developed the accurate expression of the effective stress that governs the volumetric change of saturated soils $\sigma' = \sigma - \left(1 - K/K_s\right)u_w$, where K and K_s are the volumetric change modulus of the soil skeleton and soil particles, respectively. Shen (1995), a fellow of the Chinese Academy of Science, thought,

> the two phases effective stress equations derived by Skempton from the standpoint of volumetric change- and shear strength-equivalence have already been complete. There is no need to continue the debate on this problem. Skempton's researches have adequately

illustrated that the effective stress should be calculated with the strength- or deformation-equivalence principles. It is not an objectively existed physical quantity, just like centrifugal force.

Meanwhile, with the example of deformation-equivalent effective stress, Shen derived the expression of volumetric strain increments with the increasing of total stress and pore pressure. In addition, he further derived the expression of deformation-equivalent effective stress based on the deformation-equivalence principle.

Oftentimes, equivalent stress obtained with volumetric deformation-equivalence cannot be the same to that obtained with shear strength-equivalence. That said, the volumetric deformation-equivalent "effective stress" cannot be numerically the same as the shear strength-equivalent "effective stress." The "effective stress" obtained by Skempton is only an "equivalent stress," which is a virtual physical quantity. It does not bear any common feasibility or physical basis. Therefore, it is inappropriate to equate the virtual equivalent stress to the objectively existing effective stress.

1.1.3 Effective stress of unsaturated soils

As important as the effective stress principle of saturated soils is, research on the effective stress of unsaturated soils is the most fundamental and important subject, which is the foundation and prerequisite for the other studies of unsaturated soils (Zhao *et al.*, 2013). It can mainly be classified into three categories: (1) the effective stress expressed with the single stress variable; (2) expression of the strength and deformation of soils expressed with the combination of stress state variables; and (3) determination of the effective stress with the equation of work.

(1) Effective stress expressed with the single stress variable

The success of effective stress and the effective stress equation in the saturated soils field motivated people to seek for an effective stress principle applicable for unsaturated soils. It was believed that the effective stress of saturated soils could be expanded to unsaturated soils in the early research on the effective stress of unsaturated soils where a series of effective stress or stress state variables in terms of a single variable were provided.

Biot (1941) proposed the general consolidation theory applicable for unsaturated soils containing closed air bubbles. He applied the two stress variables, effective stress $\sigma - u_w$ and pore water pressure u_w, in the constitutive equation that correlates stress with strain, and first found the necessity to distinguish the effect of total stress from that of pore water pressure. Croney *et al.* (1958) used the formula

$$\sigma' = \sigma - \beta' u_w \tag{1.1}$$

as the effective stress equation for unsaturated soils, where β' is the combination parameter representing the number of the combination points that are helpful to improve the shear strength.

Subsequently, Bishop (1959) proposed a more straightforward and widely used effective stress equation:

$$\sigma' = \sigma - u_a + \chi(u_a - u_w) \tag{1.2}$$

where χ is the parameter for stress procedure and hysteresis that depends on the degree of saturation, type of soils and dry-wet procedure, in the range of 0 to 1. When the soil body is dry, χ is 0. When the soil body is saturated, χ is 1, and Equation (1.2) degraded to the effective stress equation for saturated soils. Some researchers obtained the value of χ by controlling u_a and u_w in triaxial tests or direct shear tests. The experiment results showed that the factor χ is a function of the degree of saturation S_r,

$$\chi = g(S_r) \tag{1.3}$$

However, they did not have an explicit function that can be generally applied. To some extent, this affected the feasibility of Bishop's effective stress equation. For simplicity, many researchers suggested rewriting the equation as $\chi = S_r$.

Bishop's effective stress equation was well recognized among many geotechnical engineers for an extended period. Later, other researchers proposed the effective stress equation for unsaturated soils with a feature similar to Bishop's (Croney and Coleman, 1961; Jennings, 1960). These effective stress equations have one common property: They all contain the physical parameters of soils to consider the physical properties of soils. This gave rise to a lot of difficulties for the application of these equations in practices. Many researchers insisted that the variables describing the stress state of soils were irrelevant to the physical properties of soils. For example, the experimental results by Fung (1977) illustrated that there is no unique relationship between the soil properties obtained with tests and effective stress. In other words, the parameters' values of soil properties in Bishop's effective stress equation are different for variable problems (volumetric change or shear strength) and different stress paths. Morgenstern (1979) suggested the stress state variables should be independent, which would be combined with the characteristic parameters of soil only when the constitutive relationship for soils was constructed. In addition, the experiment results showed that the parameter χ in Bishop's equation measured with volumetric change was significantly different from that obtained with shear strength measurement.

Jennings and Burland (1962) found Bishop's effective stress did not provide a reasonable description, e.g., the collapse phenomenon of collapsing soils. They chose several types of unsaturated soils, ranged from a silty sand to a silty clay, and performed a series of oedometer and isotropic compression tests. The oedometer curves of the air-dry silt soaked at various constant applied pressures were shown in their test. These samples were dried from saturated slurry. Except for the saturated slurry, all the other samples were soaked, which were dried in different ways. This soaking behavior is obviously contrary to the effective stress theory. Based on Bishop's effective stress equation, the effective stress decreases with the gradual decreasing suction of soaked unsaturated soils. The reduction of effective stress is equivalent to unloading, which should induce the rebounding expansion of volume and increase of porosity. However, this is not what was really observed.

The results by Jennings and Burland prompted Bishop and Blight (1963) to rethink the effective stress equation for unsaturated soils. Besides the degree of saturation S_r, they thought that χ is also relevant to the other factors. They proposed a more general effective stress equation by analyzing the relationship between χ and the suction, which can be defined with the following formula:

$$\sigma' = F(\sigma - u_a, u_a - u_w) = \sigma - u_a + f(u_a - u_w) \tag{1.4}$$

Richards (1966) added the solute suction component into the effective stress equation:

$$\sigma' = (\sigma - u_a) + \chi_m (h_m + u_a) + \chi_s (h_s + u_a) \tag{1.5}$$

where χ_m is the effective stress parameter corresponding to matric suction, h_m is the matric suction, χ_s is the effective stress parameter corresponding to solute potential, and h_m is the solute suction. This equation is barely used due to its complexity.

Aitchison (1965) proposed a similar formula:

$$\sigma' = \sigma + \chi_m p''_m + \chi_s p''_s \tag{1.6}$$

where p''_m and p''_s are the matric suction and solute suction, respectively, and χ_m and χ_s are the parameters dependent on stress path, ranged from 0 to 1.

Similarly, Brackley (1971) attempted to verify the feasibility of the effective stress of unsaturated soils in describing the volumetric change of soils. The results indicated that the single-variable effective stress equation has apparent limitation in describing the volumetric change behavior of soils.

In a word, the single-variable effective stress theory of unsaturated soils was established by referring to the concept of effective stress for saturated soils from the phenomenological point of view. It is a macroscopic, visional and empirical expression, and lacks physical foundation and theoretical basis. Therefore, the approaches by the other researchers based on Bishop's equation are questionable.

(2) Representing the strength and deformation of soils by the combination of stress state variables

Since it was difficult to establish effective stress equations for unsaturated soils and researchers were questioning the appearance of the parameters of soil properties in effective stress equation, people tended to give up the concept of effective stress of unsaturated soils. They thought a single stress that controls the strength and deformation of soils, such as the effective stress of saturated soils, did not exist for unsaturated soils. Many researchers gradually adopted the concept of stress state variables, meaning the effective stress equation is divided into two independent stress state variables. The parameter χ relating to the physical properties in Bishop's effective stress was therefore not used any more.

Coleman (1967) first used $(\sigma_1 - u_a)$, $(\sigma_3 - u_a)$ and $(u_a - u_w)$ to represent the axial, confining and pore water pressure in triaxial tests, respectively. Based on these terms, a constitutive correlation of the volumetric change of unsaturated soils was established.

Bishop and Blight (1963) observed that the matric suction $(u_a - u_w)$ is different from the net stress $(\sigma - u_a)$ in its effect on changing the effective stress in their experiments. They developed a curve diagram of the correlation among porosity, matric suction and net stress. To some extent, this supported the viewpoint that matric suction and net stress correspond to different stress state variables.

Burland (1965) first clearly stated that the mechanical behavior of unsaturated soils should be relevant to two independent stress state variables: matric suction and net stress. Aitchison (1967) pointed out the complexity of the parameter χ in Bishop's effective stress equation. He believed that a specific χ in a specific stress path is composed of a specific total stress σ and matric suction $(u_a - u_w)$. Therefore, it is necessary to distinguish the effect of the total stress σ and the matric suction $(u_a - u_w)$ on unsaturated soils' behavior. Subsequently, they built up the constitutive relationship related to total stress and matric suction (Aitchison and Woodburn, 1969).

Matyas and Radhakrishna (1968) used the concept of state variables to describe the volumetric change behavior. The volumetric change of unsaturated soils is described as the three-dimensional surface related to net stress and matric suction. Barden *et al.* (1969) also suggested representing the volumetric change behavior of unsaturated soils by an independent stress combination.

Fredlund and Morgenstern (1977) should be appreciated for the wide application of two independent stress state variables theory. They conducted the stress analysis of unsaturated soils based on multiphases continuous medium mechanics. They treated an unsaturated soil as a four-phase system and assumed soil particles were uncompressible and there was no chemical reaction within soils. These assumptions are consistent with the assumptions usually applied for the soil mechanics for saturated soils. The analysis results illustrated that only two of the three stress variables, $(\sigma - u_a)$, $(\sigma - u_w)$ and $(u_a - u_w)$, are independent. The combination of any two of the variables can determine the mechanical behavior of unsaturated soils.

Usually, there are three possible combinations of stress state variables. However, the combination of net stress and matric suction is the most broadly accepted and used in general cases, mainly due to two reasons. First, in many actual situations, pore air pressure is zero, and thus the net stress and matric suction are simplified to total stress and negative pore water pressure, respectively. Second, the pore water pressure is usually negative in unsaturated soils, which is extremely difficult to measure in actual situations. By using the combination of net stress and matric suction, it could cause only one stress state variable (suction) to be undetermined, while adopting the other two combinations could lead to two stress state variables being very difficult to determine.

Subsequently, Fredlund and others conducted a great number of tests in which they increased σ, u_a and u_w by the same amount ($\Delta\sigma = \Delta u_a = \Delta u_w$) and kept the independent stress variables $(\sigma - u_a)$ and $(u_a - u_w)$ constant. The induced volumetric change of samples was zero, which confirmed the previously mentioned independent variables effective. Fredlund and Rahardjo (1993) published *Soil Mechanics for Unsaturated Soils* based on many years' research and established a more complete knowledge system of soil mechanics for unsaturated soils. After this, more and more researchers believed that the deformation and strength of unsaturated soils would be determined by two independent stress state variables.

Chen (2012) investigated the problems of choosing stress variables for unsaturated soils. He stated that the types of combination of stress state variables for unsaturated soils are more than five, instead of only one, most of which are related to the physical indexes like porosity of soils and degree of saturation, and even to the microstructures. However, the effective stress is only one of the combinations, which should be selected depending on the specific situation. Lu (2008) thought the selection of stress state variables could be random and subjective, which is determined by the type of problems investigated, such as destruction problems, elastic problems and plastic deformation problems. Alonso *et al.* (2010) reviewed and compared different selections of the stress state variables of unsaturated soils. He believed it was already an unobjectionable choice to adopt Bishop's stress (effective stress) and suction as two stress variables. He also suggested that the effect of the microstructure of unsaturated soils should be considered in Bishop's effective stress and can be reflected through the adoption of effective degree of saturation.

The two independent stress state variables theory seems to be able to interpret the regular mechanical behavior of unsaturated soils. However, its practical application incurs many difficulties. For example, the measuring process of the suction-related parameters is very complicated and requires highly delicate instruments and a long measurement period.

To indicate some more complex or specific mechanical behaviors of unsaturated soils, some researchers reused Bishop's effective stress to establish the corresponding constitutive relationship. They still kept the feature of two independent variables theory. However, different from the two independent variables theory, which is characterized with net stress and suction, they replaced the net stress in the two independent variables theory with Bishop's effective stress and retained the suction as an independent stress variable. Table 1.1 presents the stress state variables used by some researchers in their establishment of the constitutive relationship for unsaturated soils.

A trend can be recognized from Table 1.1 that Bishop's effective stress and matric suction are used as two independent stress state variables to characterize the mechanical behavior of unsaturated soils. This approach has been broadly applied and accepted. It seems that people have accepted the application of the combination of these two independent stress state variables, although they still argue about how to select the corresponding variables.

(3) Determine effective stress with the equation of work

Zhao *et al.* (2013) summarized the methods for determining effective stress with the equation of work. He thought it is a more scientific method to demonstrate the stress and deformation of unsaturated soils by use of the commonly applicable term, the deformation work in energy conservation equation, which has a solid theoretical foundation. From the perspective of deformation work, Houlsby (1997) discussed the effective stress principle of unsaturated soils and its specific expression. It is worth mentioning that the equation of effective stress of unsaturated soils based on the deformation work is not unique. Li (2007) recognized the limitation of two stress variables theory and effective stress of unsaturated soils. He considered

Table 1.1 Stress state variables used by some of the existing constitutive models

Researchers	*Stress state variables*	
Alonso *et al.* (1990) Fredlund and Rahardjo (1993) Wheeler and Sivakumar (1995) Cui and Delage (1996) Ng and Chiu (2003)	$\sigma - u_a$	$u_a - u_w$
Bolzon *et al.* (1996)	$\sigma - u_a + S_r(u_a - u_w)$	$u_a - u_w$
Modaressi and Abou-Bekr (1994)	$\sigma - \pi_c$, π_c = capillary pressure	π_c
Kohgo *et al.* (1993)	$\sigma - u_{eq}$, u_{eq} = equivalent pore pressure	$u_a - u_w - s_e$, S_e = air-entry suction
Loret and Khalili (2000, 2002)	$\sigma - u_a + \chi(u_a - u_w)$, $\chi = 1$ for $S \leq S_e$, $\chi = (Se/S)^{0.55}$ for $S > Se$	$u_a - u_w$
Gallipoli *et al.* (2003)	$\sigma - u_a + S_r(u_a - u_w)$	$\xi = f(s)(1 - S_r)$
Wheeler and Sivakumar (2003)	$\sigma - u_a + S_r(u_a - u_w)$	$n(u_a - u_w)$
Sheng *et al.* (2003)	$\sigma - mu_a$	$u_a - u_w$

the effect of gas phase in the equation of deformation work for unsaturated soils by treating air as ideal gas. By contrast, the effect of gas could not be presented in plastic deformations. Based on the energy principle and equation of work, Zhao (2009) derived the effective stress of unsaturated soils in Bishop's feature, which is a duplicate of the deformation of solid skeleton, and proposed the general effective stress principle for unsaturated soils. The nature of this general effective stress principle is that the deformation and strength of unsaturated soils cannot be uniquely determined merely by the effective stress. We must consider the effect of other general stresses at the same time, such as suction and air pressure, the corresponding general strain and the degree of saturation.

While the research on the effective stress and stress state variables of unsaturated soils has gained traction recently, the basic definition of effective stress for unsaturated soils still remains unsolved. What exactly is effective stress? Is effective stress real stress? Is the effective stress of unsaturated soils unique? No consensus has been reached regarding those issues.

1.2 Equilibrium differential equations

1.2.1 Equilibrium differential equation of foundation soil

Traditionally in soil mechanics, the estimate of stress distribution in a foundation soil can be classified into calculations of geostatic stress and additional stress. The former is caused by the self-weight of soil, and the latter is caused by external loadings (static or dynamic). In practice, the soil body was considered as a uniform semi-space infinite elastic body in the computation of geostatic stress, and a soil column was taken as an isolated body. The normal stress in a vertical direction can be directly obtained with the equilibrium condition for vertical forces. In the calculation of additional stress, the soil body is assumed to be a uniform and homogeneous semi-space infinite body. The analytical solution of additional stress under the bottom pressure (loading) of soil can be obtained with elastic theory. The additional stress of soil caused by various typical loadings, such as rectangular loading, stripe loading and circular loading, have been solved as specific elastic theoretical subjects and summarized into tables. The stress problems of bases in practice can all be transformed to the combinations of various typical loadings and solved by superimposition of stresses.

To solve the stress problems of soils with elastic theory, we need to combine the equilibrium differential equation, displacement continuous equation and the general Hook's law (i.e., linear elastic constitutive equation) in the predetermined boundary conditions. The applied equilibrium differential equation is elastic theory, which is the equilibrium differential equation of a single medium in continuum mechanics. In this case, soil is treated as a single continuous medium without the effect of pore fluid (usually pore water and pore air).

1.2.2 Biot consolidation equation

Biot (1941) derived the three-dimensional consolidation equation, which can accurately reflect the relationship between the dissipation of pore pressure and the soil skeleton deformation according to the basic equations of continuous media. He established the consolidation equation, usually called real three-dimensional consolidation theory. Directly from elastic theory, Biot's theory satisfies the equilibrium condition for soils, elastic stress-strain relationship and deformation coordination condition and considers the continuous condition of fluid. It is more theoretically strict than pseudo three-dimensional consolidation theory, yet more complicated to solve. The accurate solution can only be obtained in several conditions. Therefore, it is more widely used in finite element computation.

(1) Equilibrium equations

An infinitesimal body is taken from a soil body. If only the gravity is considered for the body, the upward direction of z axial is positive, and the compression of pressure is positive, then the three-dimensional equilibrium differential equation is:

$$\begin{cases} \dfrac{\partial \sigma_x}{\partial x} + \dfrac{\partial \tau_{xy}}{\partial y} + \dfrac{\partial \tau_{xz}}{\partial z} = 0 \\ \dfrac{\partial \tau_{xy}}{\partial x} + \dfrac{\partial \sigma_y}{\partial y} + \dfrac{\partial \tau_{yz}}{\partial z} = 0 \\ \dfrac{\partial \tau_{xz}}{\partial x} + \dfrac{\partial \tau_{yz}}{\partial y} + \dfrac{\partial \sigma_z}{\partial z} = -\gamma \end{cases} \tag{1.7}$$

where γ is the unit weight of soils, and the stress σ_x, σ_y, σ_z, τ_{xy}, τ_{yz} and τ_{xz} are total stresses.

Equation (1.7) can be rewritten as:

$$[\partial]^T \{\sigma\} = \{f\} \tag{1.8}$$

$$\text{where, } [\partial]^T = \begin{bmatrix} \dfrac{\partial}{\partial x} & 0 & 0 & 0 & \dfrac{\partial}{\partial z} & \dfrac{\partial}{\partial y} \\ 0 & \dfrac{\partial}{\partial y} & 0 & \dfrac{\partial}{\partial z} & 0 & \dfrac{\partial}{\partial x} \\ 0 & 0 & \dfrac{\partial}{\partial z} & \dfrac{\partial}{\partial y} & \dfrac{\partial}{\partial x} & 0 \end{bmatrix}, \{\sigma\} = \begin{bmatrix} \{\sigma_x\} \\ \{\sigma_y\} \\ \{\sigma_z\} \\ \tau_{yz} \\ \tau_{zx} \\ \tau_{xy} \end{bmatrix}, \{f\} = \begin{bmatrix} f_x \\ f_y \\ f_z \end{bmatrix} \text{ and } \{f\}$$

is the volumetric force in three directions.

(2) Effective stress principle

According to effective stress principle, total stress is the summation of effective stress and pore pressure u, where the pore pressure does not bear shear stress, as expressed with the following matrix:

$$\{\sigma\} = \{\sigma'\} + \{M\}u \tag{1.9}$$

where $\{M\} = [1 \quad 1 \quad 1 \quad 0 \quad 0 \quad 0]^T$.

The equilibrium equation can be written as

$$[\partial]^T \{\sigma'\} + \{M\}u = \{f\} \tag{1.10}$$

which can be expanded as

$$\begin{cases} \dfrac{\partial \sigma_x'}{\partial x} + \dfrac{\partial \tau_{xy}}{\partial y} + \dfrac{\partial \tau_{xz}}{\partial z} + \dfrac{\partial u}{\partial x} = 0 \\ \dfrac{\partial \tau_{xy}}{\partial x} + \dfrac{\partial \sigma_y'}{\partial y} + \dfrac{\partial \tau_{yz}}{\partial z} + \dfrac{\partial u}{\partial y} = 0 \\ \dfrac{\partial \tau_{xz}}{\partial x} + \dfrac{\partial \tau_{yz}}{\partial y} + \dfrac{\partial \sigma_z'}{\partial z} + \dfrac{\partial u}{\partial z} = -\gamma \end{cases} \tag{1.11}$$

where $\dfrac{\partial u}{\partial x}, \dfrac{\partial u}{\partial y}$ and $\dfrac{\partial u}{\partial z}$ are actually the unit seepage force in different directions. This equation is the equilibrium differential equation established with the free body of the soil skeleton.

(3) Constitutive equation

The physical equation of constitutive equation is known as

$$\{\sigma'\} = [D]\{\varepsilon\} \tag{1.12}$$

where the stress can be expressed by strain. Biot initially assumed the soil skeleton to be a linear elastic body, which followed the general Hook's law. Therefore, the matrix $[D]$ is a linear elastic matrix, and Equation (1.12) can be written as:

$$\begin{cases} \sigma_x' = 2G\left(\dfrac{\upsilon}{1-2\upsilon}\varepsilon_v + \varepsilon_x\right) \\ \sigma_y' = 2G\left(\dfrac{\upsilon}{1-2\upsilon}\varepsilon_v + \varepsilon_y\right) \\ \sigma_z' = 2G\left(\dfrac{\upsilon}{1-2\upsilon}\varepsilon_v + \varepsilon_z\right) \\ \tau_{yz} = G\gamma_{yz} \\ \tau_{xz} = G\gamma_{xz} \\ \tau_{xy} = G\gamma_{xy} \end{cases} \tag{1.13}$$

Where G and v are the shear modulus and Poisson's ratio, respectively.

Actually, the physical equation does not have to be limited to elastic bodies, which can also be expanded to elastic-plastic bodies, while $[D]$ is an elastic-plastic matrix.

(4) Geometry equation

Subsequently, the strain is represented in terms of displacement by applying the geometry equation. With the assumption of small deformation, the geometry equation is:

$$\{\varepsilon\} = -[\partial]\{\omega\} \tag{1.14a}$$

where $\{\omega\}=\begin{bmatrix}\omega_x\\ \omega_y\\ \omega_z\end{bmatrix}$ are the components of displacement. Expanding the Equation (1.14a) to:

$$\begin{aligned}
\varepsilon_x &= -\frac{\partial\omega_x}{\partial x},\ \gamma_{yz} = -\left(\frac{\partial\omega_y}{\partial z}+\frac{\partial\omega_z}{\partial y}\right)\\
\varepsilon_y &= -\frac{\partial\omega_y}{\partial y},\ \gamma_{xz} = -\left(\frac{\partial\omega_x}{\partial z}+\frac{\partial\omega_z}{\partial x}\right)\\
\varepsilon_z &= -\frac{\partial\omega_z}{\partial z},\ \gamma_{xy} = -\left(\frac{\partial\omega_x}{\partial y}+\frac{\partial\omega_y}{\partial x}\right)
\end{aligned} \tag{1.14b}$$

In soil mechanics, compression has a positive sign and tension has a negative sign for stress and strain. Therefore, the signs in Equation (1.14) are opposite to those in normal geometry equations of elastic mechanics.

(5) Continuous equations

The three-dimensional consolidation continuous equation is:

$$\frac{\partial\varepsilon_v}{\partial t} = -\frac{K}{\gamma_w}\nabla^2 u \tag{1.15}$$

(6) Consolidation differential equation

Substituting Equation (1.14) into (1.13), and then into (1.12), the equilibrium differential equation expressed with displacement and pore pressure are obtained, which is

$$-[\partial]^T[D][\partial]\{\omega\}+[\partial]^T\{M\}u=\{f\} \tag{1.16a}$$

For elastic-plastic problems, the expansion of this equation is complicated. Only the expanded equation for elastic problems is presented herein:

$$\begin{aligned}
&-G\nabla^2\omega_x-\frac{G}{1-2\upsilon}\frac{\partial}{\partial x}\left(\frac{\partial\omega_x}{\partial x}+\frac{\partial\omega_y}{\partial y}+\frac{\partial\omega_z}{\partial z}\right)+\frac{\partial u}{\partial x}=0\\
&-G\nabla^2\omega_y-\frac{G}{1-2\upsilon}\frac{\partial}{\partial y}\left(\frac{\partial\omega_x}{\partial x}+\frac{\partial\omega_y}{\partial y}+\frac{\partial\omega_z}{\partial z}\right)+\frac{\partial u}{\partial y}=0\\
&-G\nabla^2\omega_z-\frac{G}{1-2\upsilon}\frac{\partial}{\partial z}\left(\frac{\partial\omega_x}{\partial x}+\frac{\partial\omega_y}{\partial y}+\frac{\partial\omega_z}{\partial z}\right)+\frac{\partial u}{\partial z}=-\gamma
\end{aligned} \tag{1.16b}$$

If the volumetric strain is represented in terms of displacement using Equation (1.14), i.e., $\varepsilon_v=\{M\}^T\{\varepsilon\}=-\{M\}^T[\partial]\{\omega\}$, then Equation (1.15) becomes:

$$-\frac{\partial}{\partial t}\{M\}^T[\partial]\{\omega\}+\frac{K}{\gamma_\omega}\nabla^2 u=0 \tag{1.17}$$

whose expansion is:

$$-\frac{\partial}{\partial t}\left(\frac{\partial \omega_x}{\partial x}+\frac{\partial \omega_y}{\partial y}+\frac{\partial \omega_z}{\partial z}\right)+\frac{K}{\gamma_\omega}\nabla^2 u=0 \tag{1.18}$$

This is the continuous equation expressed with displacement and pore pressure.

The change of pore pressure and displacement with time on any point of the saturated soil body should satisfy both the equilibrium equation (1.18) and continuous equation (1.17) simultaneously. Combine these two equations and take the difference of time, and thus:

$$\begin{cases} -[\partial]^T[D][\partial]\{\omega\}+[\partial]^T\{M\}u=\{f\} \\ -\frac{\partial}{\partial t}\{M\}^T[\partial]\{\omega\}-\frac{K}{\gamma_\omega}\nabla^2 u=0 \end{cases} \tag{1.19}$$

Equation (1.19) is called Biot's consolidation equation. It contains four series of partial differential equations and four unknown variables, ω_x, ω_y, ω_z and u, which are the functions of coordinates x, y, z and time t. These four variables can be solved in specific initial conditions and boundary conditions. Equation (1.19) is a group of simultaneous equations, illustrating the coupling of the deformation equation and seepage, which is also called fluid-solid coupling. The first term of the equilibrium equation represents the force corresponding to the induced displacement; the second item represents the force corresponding to the current pore pressure. They are equilibrated to external loadings. The first term of the continuous equation represents the volumetric change corresponding to the displacement change in a unit time period; and the second term represents the seepage flux corresponding to pore pressure change. The force equilibrium and flux equilibrium are coupled, since they contain the contribution of pore pressure and deformation, respectively.

It should be particularly noted that the pore water pressure u in these equations is excess pore water pressure, i.e., the increment of pore water pressure due to the loadings.

1.2.3 Fredlund's equilibrium differential equation for unsaturated soils

Fredlund and Morgenstern (1977) and Fredlund and Rahardjo (1993) applied the method of continuum mechanics and conducted internal force equilibrium analysis on each phase in soils. They treated the surface tension film formed by the pore water in the soil skeleton as the independent fourth phase. Based on the internal force equilibrium equation, they indicated that the stress state variables that control the strength and deformation of unsaturated soils are any two combinations of the three stresses $(\sigma_t - u_a)$, $(\sigma_t - u_w)$ and $(u_a - u_w)$. The possible combinations are $(\sigma_t - u_w)$ and $(u_a - u_w)$, $(\sigma_t - u_a)$, and $(u_a - u_w)$, and $(\sigma_t - u_a)$, and $(\sigma_t - u_w)$.

(1) Equilibrium equations of unsaturated soils

The state of stress at a point in an unsaturated soil can be analyzed using a cubical element of infinitesimal dimensions. The total normal and shear stresses acting on the boundaries of the soil element are shown in Figure 1.1. The unit gravity ρg (the product of the density of soils ρ and the gravitational acceleration g) is a body force. The gravitational force acts through the centroid of the free body, which is not presented in Figure 1.1 for simplicity.

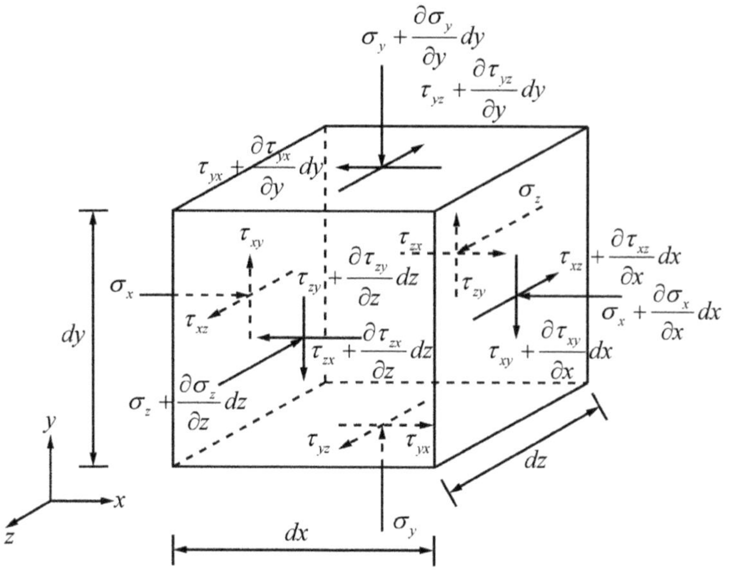

Figure 1.1 The normal stress and shear stresses on a cubical soil element of infinitesimal dimensions

The equilibrium analysis of the free body of soils was conducted based on the conservation of linear momentum. The conservation of linear momentum can be applied to the soil element in Figure 1.1 by summing force first in the y-direction:

$$\left(\frac{\partial \tau_{xy}}{\partial x} + \frac{\partial \sigma_y}{\partial y} + \frac{\partial \tau_{zy}}{\partial z} + \rho g\right) dxdydz = \left(\rho \frac{Dv_y}{Dt}\right) \tag{1.20}$$

where τ_{xy} is the shear stress on x-plane in the y-direction, σ_y is the total normal stress on y-plane, τ_{zy} is the shear stress on the z-plane in the y-direction, ρ is the total density of soils, g is gravitational acceleration, dx, dy, and dz are dimensions of the element in the x, y and z directions respectively, $\frac{Dv_y}{Dt} = \frac{\partial v_y}{\partial t} + \frac{\partial v_y}{\partial y}\frac{\partial y}{\partial t}$ is the acceleration in the y-direction, and v_y is the velocity in the y-direction.

Since the soil element does not undergo acceleration, the right-hand side of Equation (1.20) becomes zero:

$$\left(\frac{\partial \tau_{xy}}{\partial x} + \frac{\partial \sigma_y}{\partial y} + \frac{\partial \tau_{zy}}{\partial z} + \rho g\right) dxdydz = 0 \tag{1.21}$$

Equation (1.21) is commonly referred to as the equilibrium equation for the y-direction. Similarly, the counterpart in x-direction can be derived:

$$\left(\frac{\partial \sigma_x}{\partial x} + \frac{\partial \tau_{yx}}{\partial y} + \frac{\partial \tau_{zx}}{\partial z}\right) dxdydz = 0 \tag{1.22}$$

Then the equilibrium equation in the z-direction is:

$$\left(\frac{\partial \tau_{xz}}{\partial x}+\frac{\partial \tau_{yz}}{\partial y}+\frac{\partial \sigma_z}{\partial z}\right)dxdydz=0 \tag{1.23}$$

(2) Overall equilibrium

Overall equilibrium represents the force equilibrium of a complete soil element with its four phases (i.e., air, water, contractile skin, and soil particles). The equilibrium in the y-direction is analyzed as follows. The same principle is feasible on the equilibrium in the x- and z-directions. The field of total stresses in the y-direction is illustrated in Figure 1.2. The corresponding force equilibrium equation is shown in Equation (1.21).

(3) Equilibrium of independent phases

The soil particles and contractile skin are assumed to behave as solids in an unsaturated soil. In other words, it is assumed that these phases come to equilibrium under applied stress gradient. The arrangement of soil particles is referred to as the soil structure. The water phase and air phase are fluids, which flow under applied stress gradients. In the equilibrium analysis, each phase is assumed to behave as an independent, linear, continuous and coincident stress field in each direction. Therefore, an independent equilibrium equation can be written for each phase. The principle of superimposing can be applied to the equilibrium equations for each phase because the stress fields are linear. The sum of the equilibrium equations for the individual phases is equal to the overall equilibrium of the free body of soils.

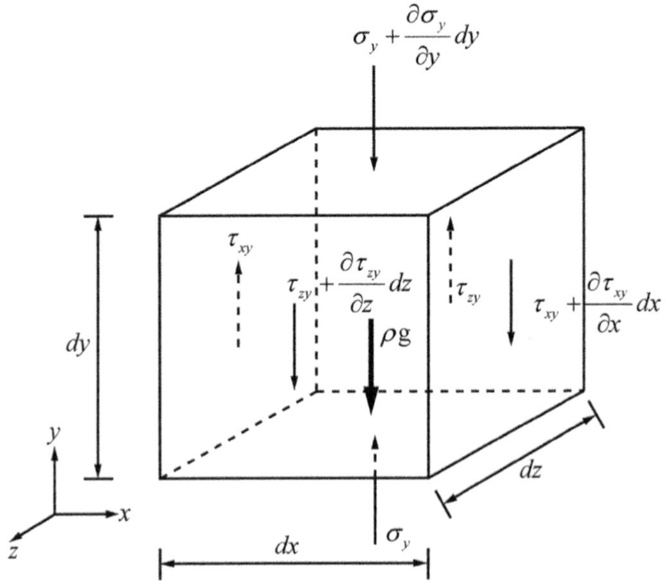

Figure 1.2 Components for the total equilibrium in the y-direction for an unsaturated soil element

The equilibrium equation for the water phase, air phase and contractile skin can be written independently. The three individual phase equilibrium equations together with the total equilibrium equation form the basis for formulating the equilibrium equation for the soil structure.

The equilibrium of water phase in the y-direction is shown in Figure 1.3. Summing forces in the y-direction gives the equilibrium equation for the water:

$$\left(n_w \frac{\partial u_w}{\partial y} + n_w \rho_w g + F_{sy}^w + F_{cy}^w \right) dxdydz = 0 \tag{1.24}$$

where u_w is the pore water pressure, F_{sy}^w is the interaction force (body force) between water phase and soil particles in the y-direction, and F_{cy}^w is the interaction force (body force) between water phase and contractile skin (body force) in the y-direction.

The equilibrium of air phase in the y-direction is shown in Figure 1.4. The summation of forces in the y-direction gives the equilibrium equation for the air phase:

$$\left(n_a \frac{\partial u_a}{\partial y} + n_a \rho_a g + F_{sy}^a + F_{cy}^a \right) dxdydz = 0 \tag{1.25}$$

where u_a is the pore air pressure, F_{sy}^a is the interaction force (body force) between air phase and soil particles in the y-direction, and F_{cy}^a is the interaction force (body force) between air phase and contractile skin in the y-direction.

Figure 1.5 presents the force components in equilibrium applied on the contractile skin in the y-direction. The variation in the normal stress associated with the contractile skin in the

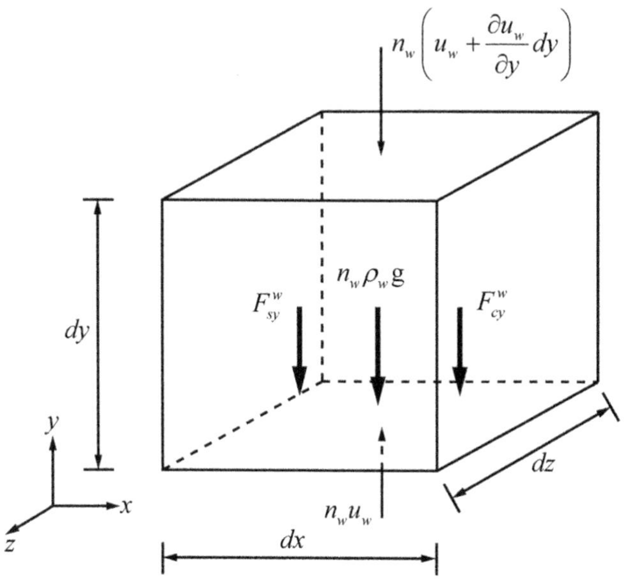

Figure 1.3 Components for force equilibrium of water phase in the y-direction.

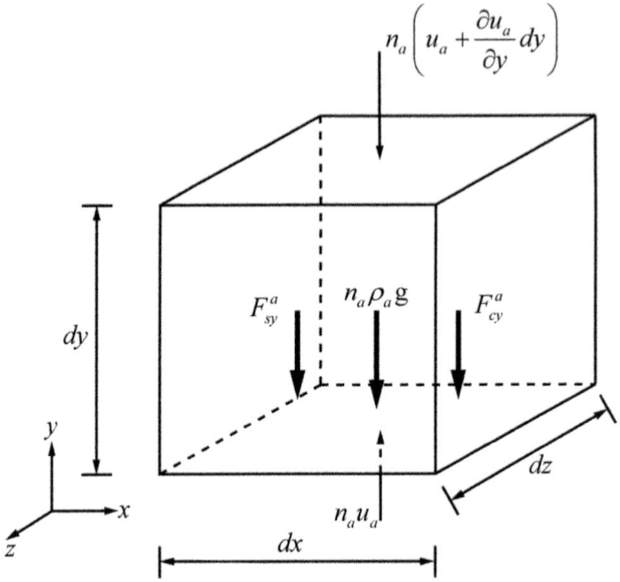

Figure 1.4 Components for force equilibrium of air phase in the *y*-direction

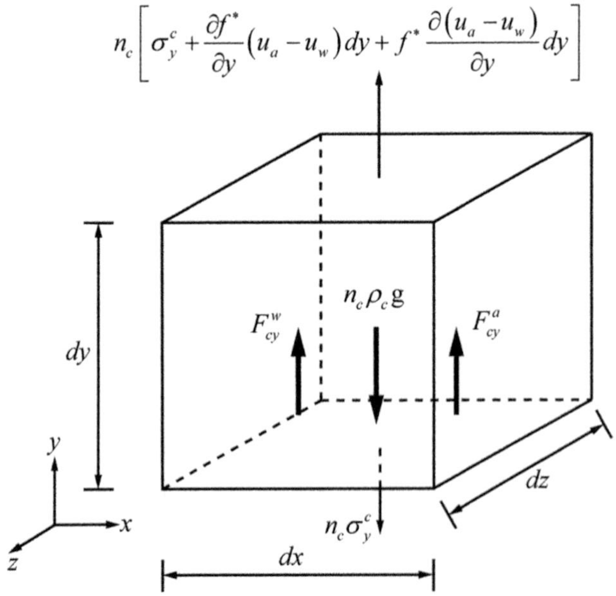

Figure 1.5 Components for force equilibrium of contractile skin in the *y*-direction

y-direction is expressed in a differential format, where σ_y^c represents the normal stress in the contractile skin on the y-plane. The equilibrium equation for the contractile skin is as follows:

$$\left\{-n_c \frac{\partial f^*}{\partial y}(u_a - u_w) - n_c f^* \frac{\partial (u_a - u_w)}{\partial y} + n_c \rho_c g - F_{cy}^w - F_{cy}^a\right\} dxdydz = 0 \tag{1.26}$$

where u_a is the pore air pressure, u_w is the pore water pressure, f^* is the ultimate interacting function between the equilibrium of contractile skin and the equilibrium of soil structure, F_{cy}^a is the interaction force (body force) between air phase and contractile skin in the y-direction, and F_{cy}^w is the interaction force (body force) between water phase and contractile skin in the y-direction.

The overall equilibrium equation for the free body of soils (i.e., Equation (1.21)) is equivalent to the resultant of the equilibrium equations for the individual phases (i.e., Equation (1.24), (1.25) and (1.26) and the equilibrium equation of soil structure). Therefore, the equilibrium equation of soil structure in the y-direction can be written as the difference between the overall equilibrium equation and the sum of the equilibrium equations for the water, air and contractile skin:

$$\begin{aligned}&\frac{\partial \tau_{xy}}{\partial x} + \frac{\partial \sigma_y}{\partial y} - n_a \frac{\partial u_a}{\partial y} - n_w \frac{\partial u_w}{\partial y} + n_c f^* \frac{\partial (u_a - u_w)}{\partial y} + \frac{\partial \tau_{zy}}{\partial z} + \\ &n_s \rho_s g - F_{sy}^w - F_{sy}^a + n_c (u_a - u_w) \frac{\partial f^*}{\partial y} = 0\end{aligned} \tag{1.27}$$

This equation can be further revised with the basic factors of pore water pressure u_w and total normal stress σ_y.

$$\begin{aligned}&\frac{\partial \tau_{xy}}{\partial x} + \frac{\partial (\sigma_y - u_w)}{\partial y} - (n_a - n_c f^*) \frac{\partial (u_a - u_w)}{\partial y} + \frac{\partial \tau_{zy}}{\partial z} + \\ &(n_c + n_s) \frac{\partial u_w}{\partial y} + n_s \rho_s g - F_{sy}^w - F_{sy}^a + n_c (u_a - u_w) \frac{\partial f^*}{\partial y} = 0\end{aligned} \tag{1.28}$$

$$\begin{aligned}&\frac{\partial \tau_{xy}}{\partial x} + (n_a - n_c f^*) \frac{\partial (\sigma_y - u_a)}{\partial y} + (n_w + n_c f^*) \frac{\partial (\sigma_y - u_w)}{\partial y} + \\ &(n_c + n_s) \frac{\partial \sigma_y}{\partial y} + n_s \rho_s g - F_{sy}^w - F_{sy}^a + n_c (u_a - u_w) \frac{\partial f^*}{\partial y} = 0\end{aligned} \tag{1.29}$$

Stress state variables can be composed with three groups of independent normal stresses taken from the equilibrium equation of soil structure (i.e., surface forces). Any two of the three stress variables (i.e., $(\sigma - u_a)$, $(\sigma - u_w)$ and $(u_a - u_w)$) can be used to describe the stress state of soil structure and contractile skin of unsaturated soils.

1.3 Continuous medium matter model for soils

Soil is a multiphase body consisting of the soil skeleton and pore fluid. The distribution of both the soil skeleton and pore fluid is discontinuous; hence soil is not a continuous medium.

To apply continuum mechanics to solve soil mechanical problems, a continuous medium matter model for soils must be established.

1.3.1 Concept of continuous medium

Matter exist in space. In mathematics, this space can be measured with the coordinates expressed with sets of real numbers. Similar to the sets of real numbers in mathematics, which are continuous sets, three-dimensional space with time is also a continuous set that can be represented with the real number set x, y, z and t. To investigate the state and movement of matters, the concept of continuous sets can be expanded to matters in continuum mechanics. In this case, matters are considered continuously distributed in space.

However, in fact, the continuous medium matters are discontinuous in micro-scale. If we shrink the investigating dimension to atomic size, the atomic core and electrons are moving ceaselessly, of which the distribution is discontinuous. Therefore, to define the physical quantity for matters in mathematical continuous space, we have to add criteria to the definition.

This can be illustrated with the definition of the materials' mass density in continuum mechanics. Assuming the space is V_0, as shown in Figure 1.6, we investigate point P in V_0 and the sub-spatial series $V_0, V_1, V_2, \ldots$ that converged at P, i.e., $V_n \subset V_{n-1}, P \in V_0$ $(n = 1, 2, \ldots)$.

Let the mass of the material in V_n be M_n, where V_n represents the volume of the subspace. Therefore, the mass density of point P of the material ρ (P) can be defined as:

$$\rho(P) = \lim_{V_0 \to 0} \frac{M_n}{V_n} \tag{1.30}$$

This type of definition is actually a mathematical abstraction. For real existed materials in nature, when the size of V_n is close to the dimension of atomic radius, this definition would encounter the following difficulties: along with the ceaseless movement of the atoms and the other basic particles of the materials, the limit of Equation (1.30) either does not exist or fluctuates with time and space.

To solve this problem, a criterion needs to be added to the limit equation (1.30). We investigate the limit of the ratio M_n/V_n. If V_n decreases and eventually approximates zero, the infinite subspace set $V_0, V_1, V_2, \ldots, V_n, \ldots$, and V_n is required to remain large enough to ensure

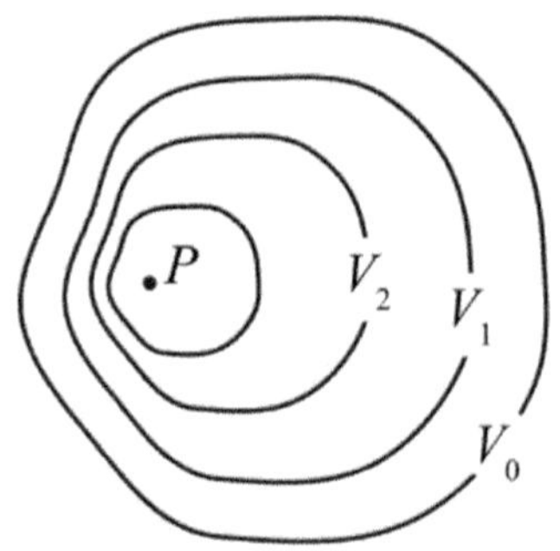

Figure 1.6 Spatial sets converged to point *P*

V_n includes an adequate number of particles. This small enough and large enough space with the matters inside is called "Representative Elementary Volume" of the matter at this point, abbreviated as *REV* and denoted by V^*. If the ratio of M_n/V_n still tends to a specific limit value in this additional criterion, ρ (*P*) is defined as the mass density of the matter at this point.

Consequently, in continuum mechanics (Feng, 1984), the definition of mass density of matters with the additional criterion is actually not the limit of the infinite sets in the subspace converged to point *P*. It is the average of the ratios between the macro mass and macro volume of the matters in the finite micro-space including point *P*. This is equalization, essentially. In other words, when the mass of matters is assumed to be uniformly distributed in microspace, we establish a mathematical model of continuous medium for real materials. The model has the mass density strictly defined with Equation (1.30) and can overcome the difficulties that may be induced in mathematical treatment during the mechanical analysis on matters.

There is no problem in applying this equalization process when the investigation is not related to microstructures of materials. Similar to mass density, the other physical quantities for materials can be defined based on this equalization.

1.3.2 Continuity of physical quantities of soils

The variables describing the physical properties and mechanical states of soils are all called physical quantities of soils. Since the spatial distribution of the soil skeleton and pores is discontinuous, the definition of physical quantities in macro conditions will encounter difficulties similar to those that continuous medium has in micro conditions. When we use Equation (1.30) to define the mass density of soils, ρ (*P*) will lose the original physical meanings once V_n tends to be smaller than the volume of skeleton particles or clusters. In this case, ρ (*P*) becomes the mass density of skeleton particles (mass points) or fluid in pores rather than the representation of the mass density of the macro soil body. This means that the limit indicated by Equation (1.30) does not exist.

Consequently, we have to add criteria in a larger space if we use Equation (1.30) to define the mass density of soils. In other words, we have to conduct equalization in a larger space containing an adequate number of skeleton particles (mass points) surrounding point *P*.

The spatial region occupied by the soil body is assumed to be V_0. For any point *P* in V_0, the subspatial set $V_0, V_1, V_2, \ldots, V_n, \ldots$ that converges to *P* is investigated, where $V_n \subset V_{n-1}$, $P \in V_0$ $(n = 1, 2, \ldots)$.

On the other hand, V^* is a finite space containing point *P*: it is small enough and tends to zero, and meanwhile large enough to include an adequate number of soil skeleton particles (mass points). V^* is called the *Representative Elementary Volume* of point *P*, i.e., *REV*.

We assume the mass of the soil skeleton contained in V_n is M_{sn}, the mass of pore water is M_{wn}, and neglect the mass of pore air. If the limit:

$$\rho(P) = \lim_{\substack{V_n \to V^* \\ V^* \to 0}} \frac{M_{sn} + M_{wn}}{V_n} \tag{1.31}$$

and

$$\rho_d(P) = \lim_{\substack{V_n \to V^* \\ V^* \to 0}} \frac{M_{sn}}{V_n} \tag{1.32}$$

exist, they are called the mass density and mass dry density of the soil body at point P, respectively.

Similarly, we call the limit:

$$\rho_s(P) = \lim_{\substack{V_n \to V^* \\ V^* \to 0}} \frac{M_{sn}}{V_{sn}} \tag{1.33}$$

the mass density of particles at point P.

Equation (1.33) has a twofold meaning: first, the mass density at point P is the average of the mass densities of the *REV*s surrounding point P; second, it should be "infinitely" equalized in *REV* to remain meaningful even approaching to point P and ensure the continuity of the definition of the density.

If the subspatial set infinitely approaches to the *REV* of point P, the limits of the other physical mechanical quantities of soils exist, and then we define these limits as the corresponding physical mechanical properties.

As shown in Figure 1.7, if all the physical mechanical state variables of any point in the space V occupied by the soil body are used to characterize the average of the *REV* at point P, which can continuously move, then these physical quantities strictly satisfy the continuous conditions in mathematics within space V. In this case, the spatial distribution continuity of the soil skeleton and pore fluid is no longer the necessary condition for the spatial continuity of physical mechanical quantities.

1.3.3 Representative Elementary Volume (REV)

Representative Elementary Volume (*REV*) has significant meaning for the definition of physical mechanical quantities and the investigation of the continuity of the quantities. However, it is usually unnecessary to define the size of the *REV*. The size can be varied for different soil bodies or different points within the same soil body. In fact, *REV* cannot be too large, otherwise the averaged value could not represent the physical quantity at point P; meanwhile, *REV* cannot be too small, because it has to contain an adequate number of pores or soil skeletal particles to ensure the possibility to obtain a meaningful statistical average. Therefore, *REV* should meet the following requirements:

(1) The average of physical quantities does not depend on the size or shape of *REV*;
(2) The average of physical quantities is continuously differentiable in space and time;

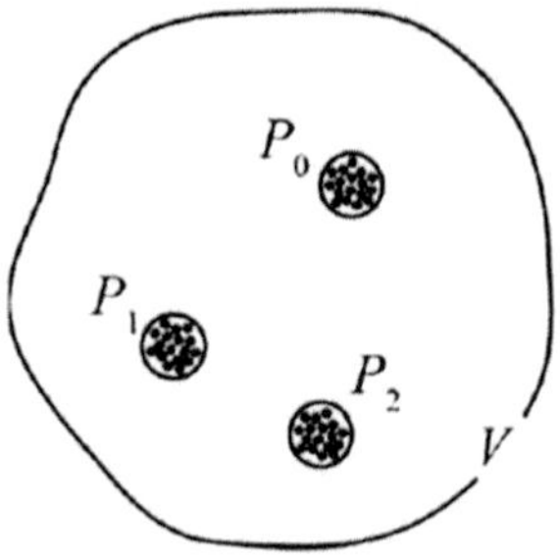

Figure 1.7 Representative Elementary Volume (*REV*)

(3) If l is the characteristic length of *REV*, and d is the characteristic length of soil skeletal particles, then:

$$l \gg d \tag{1.34}$$

(4) If L is the characteristic length in the soil area where the physical mechanical quantities change, then:

$$l \ll L \tag{1.35}$$

Actually, Equations (1.34) and (1.35) have ensured that the discontinuity of soils in micro- (meso) scale can be eliminated through the selection of *REV*, while the homogeneity or inhomogeneity in macro-scale would be unaffected.

1.3.4 Continuous medium matter model of soils

The criteria of finite volumetric space and the infinite equalization of soils' physical mechanical quantities in this space have substantially established a matter model for soils. When the approaches in macro-scale are used to solve mechanical problems of soils, the characteristic quantities describing the soil state of any point in the space, i.e., physical mechanical quantities, are all the averages in the specific area centered with this point. This area is the *REV* of this point, which is the "point" in our investigations on mechanical problems of soils in macro-scale.

If there is no concept of *REV*, the "point" in soil bodies we are discussing would either be on the soil skeleton or in pores. In that case, we could neither ensure the continuity of the physical mechanical quantities of soils nor define these quantities. For example, the soil density of one point is the mass average of all the free bodies of the soil body (the soil skeleton and pore fluid) within the *REV* of this point; the density of the soil skeleton is the mass average of the unit volume of the soil skeleton within the *REV* of this point. Furthermore, the internal forces on the unit area of the soil body within the *REV* of one point is called the stress of the soil body, while the skeleton stress at one point is the soil skeleton internal forces on the unit area within the *REV* of this point.

When the investigated point moves continuously, its corresponding *REV* moves continuously as the same time. Therefore, the physical mechanical quantities of soils strictly satisfy the mathematical continuous conditions in the continuous space occupied by the soil body, and thus continuum mechanical analysis method can be directly applied in soil mechanics.

From the perspective of matter models, the infinite equalization of soils' physical mechanical quantities in *REV* can be viewed as the infinite equalization of soil structures with the prerequisite of physical quantities remaining unchanged. In other words, the equalized soils' physical structure can keep all the physical quantities of the *REV* within infinitesimal volume. The imaginary infinitely equalized soil mass structures keeping all the physical quantities are called the continuous medium matter models of soils. In fact, to solve the soil mechanical problems, all that is required is the definition of soils' physical mechanical quantities that meet the conditions of continuous media. The continuous medium matter models of soils are not always necessary. However, these models are helpful for understanding the mechanical analysis on the infinitesimal free body of soils.

1.3.5 Pore area and soil skeleton area at cross sections of soils

As we already know, in seepage problems, the flow velocity of water through pores at cross sections of the soil body is equal to the cross sectional flow velocity divided by porosity. This actually used the concept of infinite equalization. Otherwise, we could not imagine the porosity at the cross section of the soil body is equal to its volumetric porosity. We define the pore area and skeleton area of the cross section of soils in this section, which will be used in the following chapters.

In the definition of *REV*, any cross section in the soil body we mentioned is a column corresponding to this cross section that has the thickness of *REV*. Simply speaking, the so-called cross section is a soil column with a finite thickness and the cross-sectional plane as its bottom, where this finite thickness is the characteristic length of the *REV*.

The pore area is defined herein as an example. A plane xoy is randomly taken from the soil body, as shown in Figure 1.8. An infinitesimal length δ is taken in the direction perpendicular to the plane. When δ becomes smaller but remains big enough to ensure a reasonable number of skeleton particles (mass particles) are contained in the volume formed with xoy plane and δ, the following limit is defined as the pore area of xoy plane at $z = z_0$:

$$A_v = \lim_{\substack{\Delta z \to \delta \\ \delta \to 0}} \frac{\int_{z_0}^{z_0+\Delta z} \left[\iint_A ndxdy \right] dz}{\Delta z} \tag{1.36}$$

where A_v is the pore area of xoy plane, n is the porosity, and A is the area of xoy plane.

When Δz is very small, the change of porosity is considered unrelated to z, and then Equation (1.36) becomes:

$$A_V = \iint_A ndxdy \tag{1.37}$$

Equation (1.37) is the expression of the pore area in any plane of the soil body. When n is constant:

$$A_V = nA \tag{1.38}$$

The soil skeleton area can be defined in a similar process, which is not repeated. When materials are homogeneous, the soil skeleton area is defined as:

$$A_s = (1 - n)A \tag{1.39}$$

where A_s is the area of the soil skeleton, A is the area of the soil body, and n is the porosity.

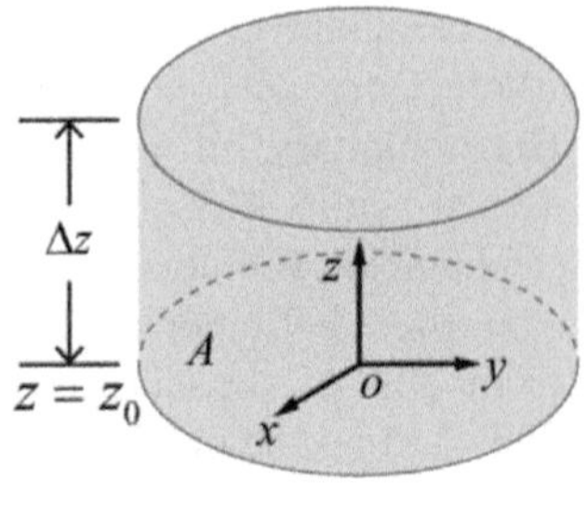

Figure 1.8 Schematic illustration for the definition of pore area

1.4 Three constitutive phases of soils

Soils are three-phase bodies consisting of solid phase, liquid phase and gas phase. The solid phase is referred to as the soil skeletons, among which there are many pores filling with fluid. The fluid is usually water or air, or the combination of these two phases. The water with the solute in it filling the pores is the liquid phase of soils, and the air with the other gas filling the pores is the gas phase of soils. The soils with all the pores completely filled with water are called saturated soils, while those with all the pores fulfilled with dry gas are called dry soils, traditionally. When there are both water and air in the pores, the soils are called unsaturated soils or wet soils. Saturated soils and dry soils are bi-phase systems, while unsaturated soils are three-phase systems.

1.4.1 Soil skeleton

The solid phase of soils includes solid particles, cementitious matters among particles and organic matters. The basic characteristics can be described with mineral composition, particle-size and particle shape. Different mineral composition, particle size distribution and shapes lead to various soil properties. For detailed explanation of the mineral composition, properties of the soil skeleton particles, cementitious properties of the solid phase in soil bodies, soil-water interactions and structures of the soil skeleton, please refer to classic soil mechanics books, such as *Fundamentals of Soil Behavior* (Mitchell, 1993). We only discuss the definition of the soil skeleton herein.

The soil skeleton can be defined as the structures consisting of the solid phase in soils that can bear and transfer loadings (Herle and Gudehus, 1999). The meaning of this definition is twofold: (1) the solid phase in soils composes the soil skeleton; and (2) this solid phase has to bear and transfer loadings. Based on this definition, the unconfined soil particles in coarse-grained soils could not be counted as a part of the soil skeleton; the pore water in fine-grained soils, which is closely related to soil particles and bears and transfers loadings with soil particles, should be treated as a constituent of the soil skeleton.

As shown in Figure 1.9, such unconfined particles exist in coarse-grained soils: although located in pores, they do not make contact with upper surrounding soil particles; except for

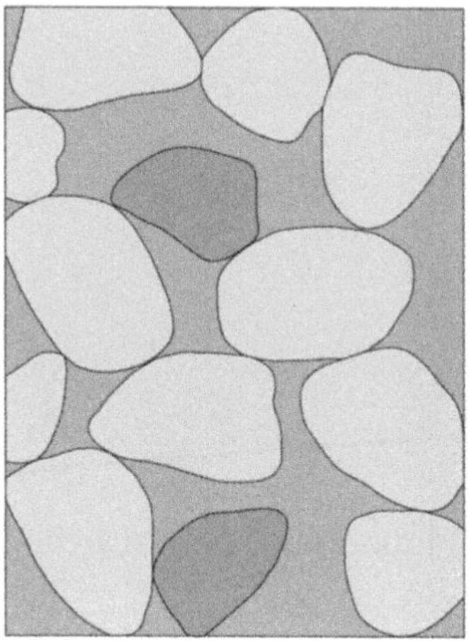

Figure 1.9 Unconfined soil particles

self-weight and pore pressure, there are no other external forces from soil boundary applied on these particles. This type of soil particle should not be treated as a constituent of the soil skeleton. The unconfining state can disappear when soil structure deforms under loading, and then these soil particles will bear and transfer loadings and become a part of the soil skeleton. Nevertheless, the number of this portion of soil particles is certainly small, and can, therefore, be neglected in solving geotechnical problems.

For clayey soils, Mitchell (1993) explicitly summarized and introduced the minerals in soil particles, the properties of water in soils and the soil-water interaction in *Fundamentals of Soil Behavior*. These studies of soils in meso-scale are significant and meaningful for understanding and mastering the nature and mechanics of soils. However, for describing the physical mechanical properties of soils in macro-scale, it appears more important to confirm whether there is a portion of water that should be treated as a constituent of the soil skeleton. Based on the existing research achievements, although we are not sure about the accurate structure of bonding water on the surface of soil particles, we positively know that it is different from the structure of normal water and ice. The existing meso-scale research results on water in soils can at least indicate that the thin film of adhesive water on the surface of soil particles was strongly absorbed by soil particles and combined with the particles tightly, which even show crystalline state. On the other hand, the studies on the water-holding characteristics of soils elucidate that soils can still retain certain water content even they were under high draining pressure (hundreds or thousands of MPa). This water content is usually called residual water content. Based on soil-water characteristic curves, residual water can be defined as a water content that would not (apparently) vary due to the increase of matric suction. Based on soil-water characteristic curves, residual water can be defined as a water content that would not (apparently) vary due to the increase of matric suction.

It can be imagined that the portion of pore water should be treated as a constituent of the soil skeleton, which bears and transfers loadings together with soil particles (as shown in Fig. 1.10). The study on the shear strength of unsaturated soils by Vanapalli *et al.* (1996) reminds us that the content of this portion of water should be the residual water content. As we will see in Chapter 3 ("Static Equilibrium Equations for Soils"), if the residual water content is assumed to be a constituent of the soil skeleton, we can directly derive the shear strength equation of unsaturated soils obtained by Vanapalli *et al.* via experiments, based on the logic that the shear strength of the soil skeleton is that of soils, which is determined by the soil skeleton stresses.

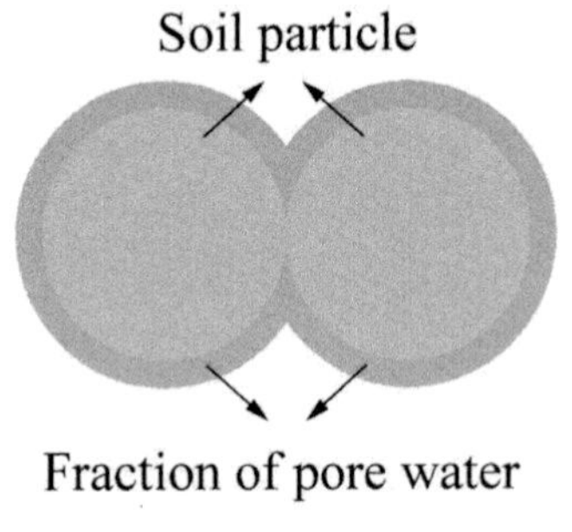

Figure 1.10 Sketch of soil particles with residual water content

1.4.2 Water in soils

The water in soils can be classified into the water in minerals and that in pores. The water in minerals, which is called mineral bonding water or crystal water, only exists in the mineral lattices of soil particles or participates in the composition of lattices. The crystal water in minerals can only be precipitated from soil particles in general. The water in soil pores is called pore water. An introduction of the knowledge on pore water can be found in the literature. Based on the nature it presents and its existing state, pore water can be classified into bonding water and unbound water.

Bonding water is the pore water tightly adhered to the particle surfaces of clay minerals due to complex physicochemical reactions. It is a decisive factor for the physical properties and mechanical behaviors of clayey soils. These complex physicochemical interactions may include hydrogen bonding, cation hydration, osmotic absorption, dipole orientation in electric field and chromatic dispersion force absorption. The strength of the interactions increases with the decrease of the distance from the soil particles' surfaces, and thus the absorption of water molecules increases. As a result, the water molecules are arranged more tightly and ordered. Meanwhile, they are less active and more steadily bonded with soil particles. With the increase of the distance, the interaction and absorption reduce, the activity increases, and the combination with soil particles is weakened.

As its properties and status vary with the equilibrium humidity (also called relative humidity or equilibrium pressure) p/p_s, bonding water can be classified into strong bonding water and weak bonding water (Curry, 1982a). When the equilibrium humidity is low, the bonding water in soils is adhered to soil particles as a constituent of the hydrate of surface layer clay, which incorporates with clay mineral lattices in a uniform ensemble. This bonding water in the hydrophilic compound on clay mineral surfaces is called strong bonding water, which belongs to solid phase. When the equilibrium humidity is relatively high, the bonding water exists as a hydration-ionization layer on the colloidal particles of clay. In this case, the surfaces of clay minerals are ionized due to the polarization of water molecules. The dissociation of the minerals on clay surface induced by water makes a portion of exchange cations leave the lattice plane and become liquid phase. This process is defined as the dissolution of ionized hydrates on the clay mineral surfaces in water. The water in the hydration-ionization layers of colloidal clay particles is called weak bonding water. It is actually the water in soils transferring from the solid phase to the liquid phase, which always continues transferring between solid and liquid phase.

Curry (1982a) suggested that the threshold equilibrium humidity for categorizing strong bonding water and weak bonding water should be 0.88. This means when $p/p_s > 0.88$, water presents as liquid and leads the ionization on clay mineral surfaces and the formation of hydration-ionization (dispersion) osmatic layers on colloidal particles. When $p/p_s \leq 0.88$, pore water exists in the status of crystal water, which belongs to solid phase.

Researchers also suggested classify bonding water into adhesive bonding water and osmotic absorbing bonding water (Curry, 1982b). Adhesive bonding water is constituted with the water molecules directly hydrated in the activation center of clay surfaces. Water molecules tightly combine to particle surfaces via pairing, static electricity and hydrogen bonding near the surfaces of soil particles. This type of water has high viscosity, high ultimate shear strength, and resistance to the effect of temperature change. With the increase of the distance from the particle surface, the adhesion gradually reduces, and water molecules form hydration under dipole-dipole interaction (van der Waal's force). Although the connection between this water layer and clay surface is weak, water molecules still arrange in order and keep high viscosity and plastic shear strength due to van der Waal's force. Osmotic absorbing bonding water is a structuralized water layer that is an interlayer from adhesive bonding water to free water. Its upper limit is equivalent to the fully developed ion-diffusing layer surrounding the clay particles, while its properties are similar to free water. However, the molecules of osmotic absorbing water cannot constitute a stable structure as free water. Similarly, the properties and status of osmotic absorbing water also vary with the distance change from the particle surfaces. The longer the distance, the weaker the hydration is, until the osmotic absorbing water directly becomes free water.

The impact factors of the properties and status of bonding water mainly include the mineral composition of clay, particle size of clay, the composition and concentration of salts dissolved in pore water, and the composition and volume of exchange cations in pore water.

Outside the bonding water, the water not affected by the surface absorption of soil particles is called unbound water, including capillary water and free water. Capillary water is a transferring type of water existing above underground water level that combined with soil particles in tiny pores of soil due to the capillary force (induced by the corporation of molecular attraction of soil particles and surface tension on the interface between water and air). Free water is also named gravity water. It can freely move and freely flow under gravity force.

The types of pore water are artificially classified. In fact, the state of water in the soil pores varies with the distance change between soil particles and transfers from bonding water to free water. The boundary between these two states should be unapparent. Since water molecules are in ceaseless moving status, there is always water molecule exchange among all the water layers., Other research results show that capillary water and bonding water cross with each other. Capillary water may appear where the weak bonding water is relatively thin, but not only at the outermost layer of weak bonding water. In other words, capillary water may have existed before the osmotic absorption bonding water fully developed (Curry, 1982c, 1982d).

To date, we still do not have the full knowledge of the structure of liquid water molecules. Only the rough structure, properties and status of the adhesive water on the extremely fine clay particles are studied. However, it is almost sure that a small amount of water in clay belongs to solid phase, and a portion of bonding water should be treated as the soil skeleton.

Soil particles have various possible arrangements in space. With the increase of the distance between soil particles, the shape of pore water changes. In the images of Figure 1.11, (a) illustrates the situation of only (strong) adhesive water skin existing at the contact point (plane) of particles; (b) is the situation when there are both strong bonding water and some weak bonding water, which may not be capillary water, at the contact plane; (c) represents the situation of more bonding water existing at the contact plane; and (d) shows that free water appears at the contact plane. Two particles can be viewed as one if they are connected without any pore water. Therefore, this situation will not be presented with diagrams.

The clay with only strong bonding water should be viewed as dry soil. This does not mean that strong bonding water skin barely affects the characteristics of soil. Just the opposite: since whether there is bonding water has significant influence on the characteristics of clay, strong bonding water should be treated as a solid phase that composes the soil skeleton together with soil particles at this time. In other words, from the perspective of stress and deformation of soils, if there is only strong bonding water at the surface of soil particles, we can consider that only the soil skeleton exists without water; similarly, if there is a portion of weak bonding water at soil particles' surface, it is also treated as a constituent of the soil skeleton.

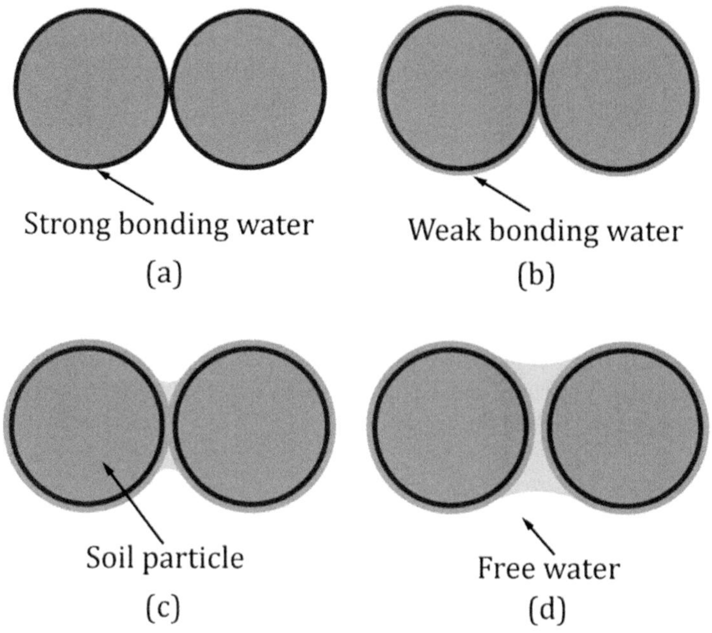

Figure 1.11 Situations of water skin existing at the contact point (plane) of particles

When the pore water content is relatively low and the soils have the characteristics of soil skeleton particles, surface tension is less likely to be present in soil pores, and so is capillary water. In reverse, the pore water in clay should be considered as a constituent of the soil skeleton when the water content is too low for capillary water to exist. In another situation, when the water content is slightly higher, capillary water – and thus surface tension – exists in soils. However, the content of capillary water is very low, which cannot be dispersed even under extremely high air pressure or eliminated by applying external forces. The capillary water bears shear strength and transfers loadings together with the soil skeleton. In this case, the pore water should be treated as a part of the soil skeleton as well. The capillary water is more like cementitious materials among soil particles, then, which should be considered as "dry soil" in mechanical analyses.

The pore water that is treated as a constituent of the soil skeleton is called soil skeleton water, which includes strong bonding water and may also include a part of weak bonding water and capillary water. The chemical composition of soil skeleton water is still water. Due to the strong absorption of it by soil skeleton particles and the capillary effect, it is tightly incorporated with soil skeleton particles to bear and transfer loadings. Increasing air pressure potential cannot disperse soil skeleton water. Therefore, the upper limit of its corresponding water content is the residual water content obtained based on soil-water characteristic curves. The soil-water characteristic curves are measured with pressure plate instruments. Increasing temperature (oven dry) or decreasing surrounding air humidity (air dry) can remove soil skeleton water. The limit for oven dry is that the water content of the soils is 0, which is hard to reach. When the actual water content of soils is lower than the aforementioned residual water content (same in the following texts), the pore water in the soils is all soil skeleton water. At this time, the water content of soils (measured with oven dry) is called the soil skeleton water content, i.e., the water content corresponding to soil skeleton water, which is abbreviated as skeleton water content.

The discussion on soil skeleton water has at least two meanings: first, it clarifies that a portion of pore water should be treated as soil skeleton water; second, it illustrates that the pore water in soils is always interconnected, even at low water content. In other words, as along as pore water exists, albeit at low water content, soil particles contact with each other with bonding water on their surfaces. Therefore, the bonding water on particles' surfaces is interconnected. There is molecule exchange between the connected bonding water, which does not need to transfer hydrostatic pressure. Only when the water content approaches or exceeds the residual water content will free water appear in the pores of soils; so will the transfer of hydrostatic pressure. Consequently, the soils with the water content lower than soil skeleton water content should be called "dry soils" in soil mechanics, as introduced before.

1.4.3 Gas in soils

Besides being filled with water, the pores in soils can also be occupied by air or the other gas. According to their existing status, the gas in soils can be classified into free gas connected to atmosphere, the gas totally closed in pore water, closed gas adhered to the surfaces of soil particles and the gas dissolved in pore water.

In most cases, the gas in soils is air. The free gas connected to the atmosphere outside is prone to overflow, which has little effect on the properties of soils. The gas dissolved in pore water is generally invisible with the naked eye and has no effect on soils' properties except increasing the compressibility of pore water. The closed gas in pore water and that adhered to the surfaces of soil particles normally exist as visible gas bubbles. According to the definition of *REV*, when the diameter of gas bubbles is large enough to reach a specific size, e.g., the maximum particle size of soil skeleton particles or the characteristic size of *REV*, the surface of the gas bubbles should be treated as the boundary of soils. The existence of closed gas bubbles increases the compressibility of water. The former (the closed gas bubbles) may destroy the original soil structure when it is disturbed and overflows, while the latter (the gas adhered to the surfaces of soil particles) can affect the effective stress of soils (Chapter 3). The volume of closed gas bubbles changes with the variation of pressure. The volume decreases with the increase of pressure, and vice versa. Therefore, the existence of closed gas can influence the deformation of soils. On the other hand, the closed gas bubbles may clog the seepage path in soils and reduce soils' permeability.

In addition, in organic soils, such as sludge and peat soils, due to the decomposition of microorganisms, there can be some toxic gas and flammable gas gathered in soils, including CO_2, H_2S and methane. Among these gases, CO_2 has the strongest absorption. It is buried in deeper soil layers, of which the concentration increases with the increase of depth.

1.4.4 Definition of physical indexes for soils' three phases with the consideration of soil skeleton water

The definition of physical indexes for soils' three phases without considering soil skeleton water are available in all the soil mechanics textbooks, which are summarized in Table 1.2.

Table 1.2 Physical indexes for soils' three phases and conversion between them

Index	*Definition*	*Common conversion equations*	*Unit*
Natural density ρ	$\rho = \frac{m}{V}$	$\rho = \rho_d(1+w)$, $\rho = \frac{\rho_s(1+w)}{1+e}$	kg/m^3
Density of soil particles ρ_s	$\rho_s = \frac{m_s}{V_s}$	$\rho_s = \frac{Se}{w}\rho_w$	kg/m^3
Dry density ρ_d	$\rho_d = \frac{m_s}{V}$	$\rho_d = \frac{\rho}{1+w}$, $\rho_d = \frac{\rho_s}{1+e}$	kg/m^3
Saturated density ρ_{sat}	$\rho_{sat} = \frac{m_s + V_v\rho_w}{V}$	$\rho_{sat} = \rho_d + n\rho_w$, $\rho_{sat} = \frac{\rho_s + e\rho_w}{1+e}$	kg/m^3
Buoyant unit weight Υ'	$\gamma' = \gamma_{sat} - \gamma_w$	$\gamma' = \frac{\rho_s - \rho_w}{1+e} \cdot g$, $\gamma' = \frac{(\rho_s - \rho_w) \cdot \rho}{\rho_s(1+w)} \cdot g$	kN/m^3
Void ratio e	$e = \frac{V_v}{V_s}$	$e = \frac{\rho_s(1+w)}{\rho} - 1$, $e = \frac{\rho_s}{\rho_d} - 1$	—
Porosity n	$n = \frac{V_v}{V_s} \times 100\%$	$n = \frac{e}{1+e}$, $n = \left(1 - \frac{\rho}{\rho_s(1+w)}\right) \times 100\%$	%
Volumetric water content θ	$\theta = \frac{V_v}{V_s} \times 100\%$	$\theta = \frac{Se}{\rho_s}\rho_w\rho$, $\theta = \left(\frac{\rho}{\rho_d} - 1\right)\rho$	%
Mass water content W	$w = \frac{m_w}{m_s} \times 100\%$	$w = \frac{Se}{\rho_s}\rho_w$, $w = \frac{\rho}{\rho_d} - 1$	%
Degree of saturation S	$S = \frac{V_v}{V_s} \times 100\%$	$S = \frac{w \cdot \rho_s}{e \cdot \rho_w} \times 100\%$	%

In the definitions of the physical indexes for soils' three phases in Table 1.2, soil skeleton water is not considered as a constituent of the soil skeleton. It does not affect the natural density, soil particle density, saturated density and buoyant density regardless if the existence of soil skeleton water is considered, while it does influence the water content, porosity, void ratio, degree of saturation and dry density. To clearly distinguish these indexes, we call them – without considering the soil skeleton water – absolute indexes. For example, considering the water content without considering the soil skeleton water as one constituent of the soil skeleton is called absolute water content and still abbreviated as water content. Similarly, porosity, void ratio and degree of saturation are all meant the corresponding three phases indexes without considering soil skeleton water when no specific interpretation is given.

When the soil skeleton water is treated as a constituent of the soil skeleton, the definitions of porosity, void ratio, water content and degree of saturation of soils need to be modified. Before presenting the relevant equations and definitions, we will discuss residual water content and skeleton water content.

(1) Residual water content

As mentioned before, when the increase of matric suction does not induce any (apparent) variation, the water content of soils is called residual water content, and the corresponding mass water content and volumetric water content are denoted with W_r and θ_r. The residual water content is the content of pore water that cannot be dispersed by air pressure, which can be determined with a soil-water characteristic curve. Usually, the water content of soils corresponding to the matric suction tending to "infinite large" in soil-water characteristic curves is used as the residual water content. It is used as the upper limit of the soil skeleton water content of soils. When the water content of soils is higher than the residual water content, the skeleton water content is equal to the residual water content. Conversely, when soils' water content is lower than the residual water content, the skeleton water content is equal to the absolute water content.

(2) Skeleton water content

The content of the pore water, which exists as a constituent of the soil skeleton, is called skeleton water content. The corresponding mass water content and volumetric water content are denoted with w_s and θ_s. As we introduced earlier, the upper limit of skeleton water content is residual water content. When soils' water content is lower than the residual water content, the skeleton water content is equal to the absolute water content of soils.

(3) Effective water content

The water content of the soils, which does not consider soil skeleton water, is called effective water content. The corresponding mass water content and volumetric water content are denoted with w_e and θ_e. Effective water content is equal to the difference between the water content and skeleton water content. When the absolute water content of soils is higher than the residual water content, the effective water content can be calculated using the absolute water content to subtract the residual water content; on the other hand, when the absolute water content of soils is lower than the residual water content, the effective water content is 0, and the soil skeleton water content is equal to the absolute water content, which can be illustrated with the following equation:

$$W_e = W - W_r \text{ and } \theta_e = \theta - \theta_r.$$

(4) Effective degree of saturation

When the soil skeleton water is considered, the definition of degree of saturation becomes relatively complicated, which includes degree of saturation, effective degree of saturation and residual degree of saturation. Degree of saturation is the one corresponding to absolute water content, denoted with S; effective degree of saturation corresponds to the effective water content, denoted with S_e; the residual water content corresponds to the residual degree of saturation, denoted with S_r. Therefore, effective degree of saturation can be expressed as:

$$S_e = \frac{S - S_r}{1 - S_r} \tag{1.40}$$

where $S_r = \frac{n_r}{n}$ is the residual degree of saturation (n_r is the residual porosity; the definition is shown as follows). Similarly, the effective degree of saturation is equal to 0 when soils' water content is lower than the residual water content.

(5) Effective porosity

When the soil skeleton water is out of consideration, the porosity of soils is the ratio of the entire pore volume and the total volume of soils (please revisit the definition of *REV*). When the skeleton water is considered as a constituent of the soil skeleton, the traditional pore volume (e.g., the pore volume measured with oven dry method) decreases. The porosity corresponding to skeleton water is called skeleton water porosity and denoted with n_{sw}. Thus:

$$n_{sw} = \frac{V_{sw}}{V} \tag{1.41}$$

where V_{sw} is the volume occupied by skeleton water and V is the volume of soils.

The ratio of pore volume and total volume is called effective porosity when the skeleton water is treated as the soil skeleton rather than pores. It is denoted with n_e and expressed as:

$$n_e = \frac{V_v - V_{sw}}{V} = n - n_{sw} \tag{1.42}$$

Where V_v is the pore volume including skeleton water, which is the aforementioned total pore volume.

The proportion of the pores corresponding to the irremovable water under air pressure is called residual porosity, which is denoted with n_r and expressed with:

$$n_r = \frac{V_r}{V} \tag{1.43}$$

where V_r is the volume corresponding to residual water content.

Since we consider the residual water content as the upper limit of soil skeleton water content,

$$n_{sw} \leq n_r \tag{1.44}$$

For unsaturated soils, pores are filled with both water and air. The porosity corresponding to pore water is denoted with n_w, and pore air corresponded one is denoted with n_a, then:

$$n = n_w + n_a \tag{1.45}$$

Deducting the skeleton water from the pore water of unsaturated soils, the rest of the porosity corresponding to pore water is called the pore water-corresponded effective porosity, abbreviated as effective porosity and denoted with n_{ew}. Therefore, we have

$$n_{ew} = n_w - n_r \tag{1.46}$$

When the water content of soils is lower than the residual water content, the corresponding effective porosity is 0.

(6) Effective void ratio

Similar to porosity, if the soil skeleton water is not considered, the ratio between the total pore volume of soils and the volume of soil particles is the void ratio. When the skeleton water is treated as a constituent of the soil skeleton, the pore volume will reduce and the volume of soil particles will increase. The void ratio corresponding to soil skeleton water is called skeleton water void ratio, denoted with e_{sw}, and then:

$$e_{sw} = \frac{V_{sw}}{V - V_v + V_{sw}} \tag{1.47}$$

which is

$$e_{sw} = \frac{n_{sw}}{1 - n_e} \tag{1.48}$$

When the skeleton water is treated as the soil skeleton, the void ratio of soils is called effective void ratio, denoted with e_e, and:

$$e_e = \frac{V_v - V_{sw}}{V - V_v + V_{sw}}, \quad e_{sw} = \frac{n - n_{sw}}{1 - n_e} \tag{1.49}$$

(7) Effective dry density

When skeleton water is considered as a constituent of the soil skeleton, the dry density of soils will increase. The mass of soil particles that includes soil skeleton water in the unit volume is called the effective dry density of soils, of which is definition is

$$\rho_d' = \frac{m_s + m_{sw}}{V}, \quad \rho_d' = \rho_d + n_{sw}\rho_w \tag{1.50}$$

where ρ_d' is the dry density of soils considering the soil skeleton water, i.e., effective dry density; ρ_d is the conventional dry density of soils; ρ_w is the density of water; m_s is the mass of soil particles; m_{sw} is the mass of soil skeleton water; and V is the soil volume.

The definition of physical indexes for soils' three phases considering soil skeleton water is summarized in Table 1.3.

1.5 Soil-water potential and its components

The soil-water potential at one point in soil is the potential energy possessed by the unit amount of water in the Representative Elementary Volume (*REV*) of this point. It is measured with the work that the unit amount of pore water moves or changes from standard reference status to current status. If the environment works on pore water in moving or changing process, the soil-water potential in this status is positive; if pore water works on the environment, the potential is negative. The soil-water potential is numerically equal to the applied work. The soil-water potential can also be defined from another point of view, which is the work applied by the moving pore water: during the movement of the unit amount of pore water from certain status to the standard reference status, if the environment works on pore water, the soil-water potential is negative; if pore water works on the environment, the potential is positive.

Soil-water potential is used to describe the status of pore water in soils. The soil-water potential of one point is determined by the difference between the potential energy of the water at this point in the current status and the potential energy in standard reference status. Usually, pure free water at a specific height and temperature that bears standard atmospheric pressure is taken as the standard reference status.

Table 1.3 Physical indexes for soils' three phases considering soil skeleton water and conversion between them

Index	*Definition*	*Common conversion equations*	*Unit*
Natural density ρ	$\rho=\frac{m}{V}$	$\rho=\rho_d(1+w)$, $\rho=\frac{\rho_s(1+w)}{1+e}$	kg/m^3
Density of soil particles ρ_s	$\rho_s=\frac{m_s}{V_s+V_{sw}}$	$\rho_s=\frac{S_e e}{w_e}\rho_w$	kg/m^3
Effective dry density ρ_d'	$\rho_d'=\frac{m_s+m_{sw}}{V}$	$\rho_d=\frac{\rho}{1+w_e}$, $\rho_d=\frac{\rho_s}{1+e_e}$	kg/m^3
Saturated density ρ_{sat}	$\rho_{sat}=\frac{m_s+V_v\rho_w}{V}$	$\rho_{sat}=\rho_d+n\rho_w$, $\rho_{sat}=\frac{\rho_s+e\rho_w}{1+e}$	kg/m^3
Buoyant unit weight γ'	$\gamma'=\gamma_{sat}-\gamma_w$	$\gamma'=\frac{\rho_s-\rho_w}{1+e}\cdot g$, $\gamma'=\frac{(\rho_s-\rho_w)\rho}{\rho_s(1+w)}\cdot g$	kN/m^3
Effective void ratio e_e	$e_e=\frac{V_v-V_{sw}}{V-V_v+V_{sw}}$	$e=\frac{\rho_s(1+w)}{\rho}-1$, $e=\frac{\rho_s}{\rho_d'}-1$	-
Effective porosity n_e	$n_e=\frac{V_v-V_{sw}}{V}\times 100\%$	$n_e=\frac{e_e}{1+e_e}$, $n_e=\left(1-\frac{\rho}{\rho_s(1+w_e)}\right)\times 100\%$	%
Effective volumetric water content θ_e	$\theta_e=\frac{V_v-V_{sw}}{V_s+V_{sw}}\times 100\%$	$\theta_e=\frac{S_e e_e}{\rho_s}\rho_w\rho$, $\theta_e=\left(\frac{\rho}{\rho_d'}-1\right)\rho$	%
Effective mass water content w_e	$w_e=\frac{m_w-m_{sw}}{m_s+m_{sw}}\times 100\%$	$w_e=\frac{S_e e_e}{\rho_s}\rho_w$, $w_e=\frac{\rho}{\rho_d'}-1$	%
Effective degree of saturation S_e	$S_e=\frac{V_v-V_{sw}}{V_s+V_{sw}}\times 100\%$	$S_e=\frac{w_e\cdot\rho_s}{e_e\cdot\rho_w}\times 100\%$	%

Total soil-water potential (abbreviated as total water potential) includes gravity potential, pressure potential, matric potential, solute potential and temperature potential, which are introduced as follows:

(1) Gravity potential ψ_g

Gravity potential is caused by the existence of gravity field and determined by the height or vertical location of the water in the soils. During the water's moving from one point to the plane of standard reference status with the other terms kept constant, the work applied by the water in soils is the gravity potential of this point.

The reference plane can be randomly selected, usually at ground surface or underground water surface. The origin of vertical coordinate axial z is set on the reference surface. The upward or downward direction can be selected as the positive direction for axial z depending on the needs. When the vertical coordinate is selected, the gravity potential, E_g, of the water with the mass of M at coordinate z in soils is:

$$E_g = \pm M g z \tag{1.51}$$

where g is gravitational acceleration. When the upward direction of z axial is positive, it takes "+" sign for Equation (1.51); when the downward direction of z axial is positive, it takes "–" sign for Equation (1.51). Obviously, all the points and gravity potentials above the reference plane are positive, and the ones under the plane are negative. Based on Equation (1.51), the gravity potential of the unit mass of water is:

$$\psi_g = \pm g z \tag{1.52}$$

The gravity potential of the unit volume of water is

$$\psi_g = \pm \rho_w g z \tag{1.53}$$

The gravity potential of the unit weight of water is

$$\psi_g = \pm z \tag{1.54}$$

(2) Pressure potential ψ_p

Pressure potential is caused by the pressure difference in a pressure field. The defined pressure in standard reference status is standard atmospheric pressure or localized pressure. If the pressure of the water on one point is different from the atmospheric pressure in reference status, then an additional pressure Δp is considered to exist at this point. During the process in which the unit amount of water moves from this point to standard reference status, with the other terms remaining constant, the work applied by water only due to the additional pressure is called the pressure potential of this point. When the volume of water is V, the water with the pressure difference or additional pressure Δp has the pressure potential E_p, as:

$$E_p = V \Delta p \tag{1.55}$$

For saturated soils, the additional pressure at the depth h of underground water is ρ_w g h. Therefore, the pressure potential of the unit mass of water at this point is

$$\psi_p = g h \tag{1.56}$$

The gravity potential of the unit volume of water is

$$\psi_p = \rho_w g h \tag{1.57}$$

The gravity potential of the unit weight of water is

$$\psi_p = h \tag{1.58}$$

Consequently, the pressure potential is $\psi_p \geq 0$ for saturated soils.

For unsaturated soils, considering the connectivity of open pores, the pressure on each point is atmospheric pressure. Therefore, the additional pressure on each point is $\Delta p = 0$, and hence the corresponding pressure potential $\psi_p = 0$. However, when there are unfilled closed pores in unsaturated soils, the pressure equilibrated with water phase may be different from the atmospheric pressure. The derivative pressure potential is called air pressure potential. The existence of closed air bubbles and corresponding air pressure potential has certain influence on the water in soils, which is worth further investigation. However, this has not been considered in current studies. In addition, in laboratory conditions, when unsaturated soil samples are placed in a closed container in which the pressure is higher than the atmospheric pressure (pressure chamber), water has a positive pressure potential.

(3) Matric potential ψ_m

The matric potential of the water in soils is caused by the suction on the water in soils by the solid particle matric in soils. The suction mechanism is very complicated; however, it can be classified into suction effect and capillary effect. Free water is used as the standard for the reference status, which does not include the effect of solid particle matric. The unit amount of water is moved from one point in unsaturated soils to the standard reference status, during which all the other terms remain constant except for the effect of soil particle matric. In this case, the work applied by water is the matric potential of this point. Actually, this work is a negative value. This is because that the water has to work to resist the suction effect by solid particle matric in the realization of the previously mentioned movement. It can be recognized that the matric potential of the water in unsaturated soils is always negative, i.e., $\psi_m < 0$, and the matric potential of the water in saturated soils is $\psi_m = 0$.

The quantity of the suction effect on the moisture by solid particle matric is related to the water content in soils. Therefore, the matric potential ψ_m of the water in unsaturated soils is a function of the water content of soils θ.

Due to the complexity of the suction effect of solid particle matric on the moisture, a quantified relation of the matric potential cannot be determined theoretically. It can only be measured outside or in a laboratory by referring to the testing methods in relevant literature. The measurement principle is indicated with the tensiometer. As shown in Figure 1.12, the tensiometer is constituted with a porous ceramic head and a U-shaped tube. The tubes in the meter are filled with water, and the ceramic head is placed in the measured point in soils. When the water surface in U-shaped tube is steady, the water potential in the negative pressure meter and that in the soils around the ceramic head are equilibrated. It illustrates that the soil-water potential at point A, ψ_A is equal to the water potential at point B, ψ_B. Since there is no difference in the concentrations of solute and temperatures at these two places, the solute potential and temperature potential at point A are the same as their counterparts at point B, respectively. Taking the water surface at point B as the reference plane, the gravity potential, pressure potential, matrix potential and total water potential of these two points are:

$\psi_{Ag} = H$, $\psi_{Ap} = 0$, $\psi_{Am} = \psi_{Am}$, and $\psi_A = H + \psi_{Am}$

and $\psi_{Bg} = 0$, $\psi_{Bp} = 0$, $\psi_{Bm} = 0$, and $\psi_B = 0$.

Since $\psi_A = \psi_B$, the matric potential of point A is $\psi_{Am} = -H$.

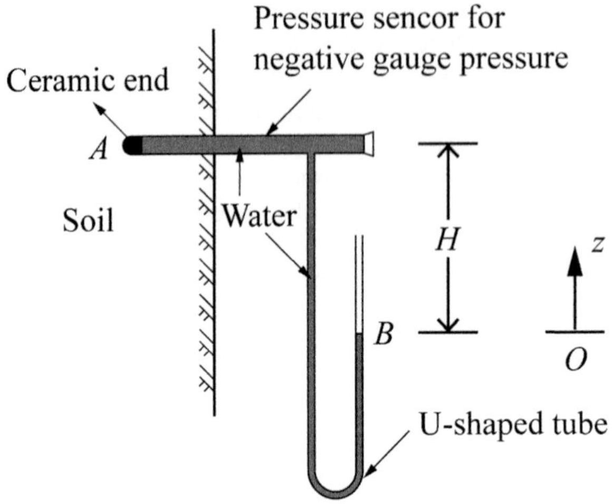

Figure 1.12 Schematic diagram of a tensiometer

The mechanism of positive pressure potential ψ_p is essentially different from that of negative matric potential ψ_m. However, these potentials sometimes are unified to facilitate the problem analysis. Then matric potential is called negative pressure potential. In this case, if the pressure potential ψ_p is expressed with pressure water head h (>0), the matric potential ψ_m is represented with negative pressure water head h_m (<0) or h (<0). This unification benefits the analysis on saturated-unsaturated flow appreciably.

(4) Solute potential ψ_s

The existence of solute potential can be proved with penetration tests. As shown in Figure 1.13, pure water is on the left side, sugar water (or the other solution) is on the right side and an ideal semipermeable membrane (a type of membrane that allows water molecules, but not solute molecules, to pass through) is in the middle. The water molecule on any side can move to the other side through the membrane. However, due to the absorption effect of the solute molecules on water, the eventual effect is that water molecules in the left tube keep entering the right tube. This leads the liquid level to keep rising until a height difference h forms between the liquid levels on the two sides. This diffusion effect in a single direction is called permeation in chemistry. The liquid level difference between the two tubes can be viewed as the pressure that has to be applied on the solution surface to prevent water from penetration. This pressure is the pressure difference between the two sides of the semi-permeable membrane before the permeation effect, which is called the permeation pressure of the original solution. The fact of water molecules transferring from the pure water to the solution through the semi-permeable membrane illustrates that the existence of solutes lowers the potential energy of water. The permeation potential P_s can be written as:

$$P_s = \frac{c}{\mu} RT \tag{1.59}$$

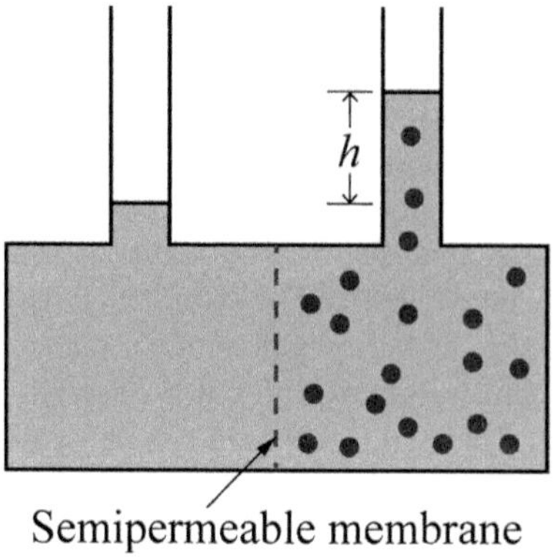

Figure 1.13 Schematic diagram of penetration test

where c is the solute mass (g/cm^3) in the unit volume of solution, which is usually called the concentration of the solution. In the solution with the volume of V (cm^3), if the solute mass is M_s (g), then $c = M_s/V$. The molar mass of the solute is μ (g/mol), numerically equal to the molecular weight of the solute. Therefore, c/μ is the concentration of the solution in molar (mol/cm^3). T is the thermal mechanical temperature (K), R is the molar gas constant, sometimes called the general gas constant. When the unit of permeation pressure P_s is Pa, $\mathrm{R} = 8.31\times10^6$ Pa·cm^3/(mol·K).

According to the above description, the solute potential of the unit volume of soil water that contains certain solute ψ_s is

$$\psi_s = -\frac{c}{\mu}\mathrm{R}T \tag{1.60}$$

Generally, there is no semi-permeable membrane in soils. The existence of the solute in water does not apparently affect the movement of the water molecules in soils. Therefore, the influence of solute potential is not considered.

(5) Temperature potential ψ_T

Temperature potential is caused by the temperature difference in temperature fields. The temperature potential of the water at any point in soils is determined by the difference between the temperature of this point and the temperature at standard reference status. Based on Equation (1.60), temperature potential can be expressed with $\psi_T = -S_e\Delta T$, where S_e is the entropy of the unit amount of water. The movement flux of water caused by temperature difference is relatively small. Therefore, the effect of temperature potential is always neglected in the analysis of the water's movement in soils.

The distribution and change of the temperature in soils affect the movement of the water in soils in many aspects. Some of these influences largely exceed the effect of temperature potential itself. For example, according to its effect on the physicochemical properties (e.g., viscosity, surface tension and osmotic pressure) of water, temperature affects the quantity of matric potential and solute potential and the movement factors of the water in soils. Temperature status also determines the phase change of soils, and the water status in soils determines the thermal characteristic factors to a large degree. The phase change of water will become an important factor if it happens. Therefore, the crossing and combination of the thermal movements of the water in soils are more important for real problems.

The components of soil-water potential are not equally significant for practical problems. As aforementioned, in the analysis of the motions of water, solute potential and temperature potential can usually be neglected. For saturated soils, the total water potential ψ is composed of pressure potential ψ_p and gravity potential ψ_g, since $\psi_m = 0$. If the water potential is expressed with that of the unit weight of water, it is generally called water head. For saturated soils, total water potential or total water head can be written as:

$$\psi = h \pm z \tag{1.61}$$

where h is the pressure head, of which the value is the depth under the underground water level; z is the position head, of which the sign is determined by the direction of axial z.

For unsaturated soils, the pressure potential is $\psi_p = 0$ without considering the air pressure potential. Therefore, the total water potential ψ is composed of matric potential ψ_m and gravity potential ψ_g, i.e.:

$$\psi = \psi_m \pm z \tag{1.62}$$

Similarly, if it is expressed with water head, the total water head in Equation (1.62) is composed of negative pressure head $h = \psi_m$ and position head.

For the convenience of introduction, the soil-water potential without gravity potential is called water potential, denoted with ψ_w; the positive water potential of the unit volume of pore water in unsaturated soils is called soil suction or total suction, denoted with s_w; and the soil suction corresponding to matric potential and solute potential are called matric suction and solute suction, respectively, denoted with s_m and s_s. In other words, matric suction is negative matric potential; and solute suction is negative solute potential. Total suction is the negative water potential, which is equal to the sum of matric suction and solute suction:

$$s_w = s_m + s_s = -\psi_w = -(\psi_m + \psi_s) \tag{1.63}$$

1.6 Soil-water characteristic curves

For water in soils, its amount is represented with water content or degree of saturation, and its energy is represented with soil-water potential. In static equilibrium condition, the relationship between the energy and the amount of the water in soils is the soil-water characteristic curve (SWCC). SWCC needs to be obtained via experiments. It is a native property, i.e., a constitutive relationship, which indicates the water-holding capacity of soils.

1.6.1 Physical meaning and basic characteristics of SWCC

The definition of the soil-water characteristic curve (SWCC) was first developed in agrology and soil physics. It is used to describe the change of matric suction in soils with water content, and thus also called soil-water retention curve (SWRC or WRC). According to its measuring approach, SWCC reflects the relationship between the potential energy of a unit amount of pore water in static condition and water content in the soils at a determined state. The determined state herein means determined mineral composition of soil particles, particle-size composition, density or void ratio, soil skeleton structure, stress status, and stress history. The testing results of SWCC showed that water content is positively correlated to water potential. That said, the lower the water content, the more the work is applied on discharging the water at this content in soils, and vice versa.

SWCC can also be understood from the perspective of the water retaining effect of soils on water. When soils are in saturated state, the water content is saturated water content, and the suction is 0. If a tiny suction is applied on the soil, there will be no water discharged from the soil, which will still remain saturated. Only when the suction increases to a threshold value do soils start to drain, since the maximum pore in soils cannot resist the suction to retain moisture. When the suction further increases, the submaximum pore continues to drain and the water content of soils further decreases. While the suction keeps increasing, the soils keep draining from large pores to small pores. Therefore, the water content becomes lower and lower, and the soil pores that retain moisture are smaller and smaller. When the suction is very high, only extremely narrow pores can hold extremely limited moisture. Consequently, with the increase of suction, water draining can barely be observed, and the water content hardly changes with the increase of suction.

The above phenomena indicate that the larger soil pores have a smaller retaining force on water, while smaller soil pores have a larger retaining force on water. The process that soils experience from saturation to the beginning of draining means that air begins to enter the maximum soil pores. The suction at this time is named air-entry value. The air-entry value is called displacement pressure in the petroleum industry and bubbling pressure in the ceramic industry. As illustrated in Figure 1.14, the air-entry value of coarse sandy soils and well-structured soils is relatively small, while that of fine clay is relatively large. Furthermore, since sandy soils have relatively uniform pore size, the appearance of their air-entry value is usually more appreciable than fine particle soils. The water content that will not change due to the increase of matric suction any further is called residual water content. Residual water content is the water content of soils that cannot be discharged no matter how the matric suction increases. The air-entry value and residual water content need to be measured with experiments.

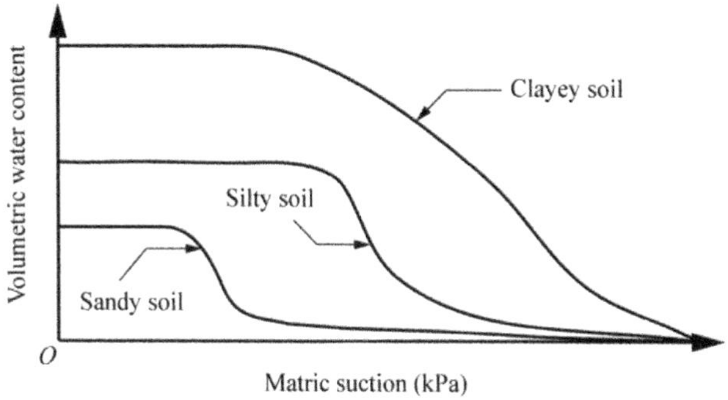

Figure 1.14 Soil-water characteristic curves of sandy soil, silty soil and clayey soil

The existence of air-entry value is initiated by capillary effect, which is the surface tension film in soil pores. There would be no water discharged from soils when the applied air pressure is not high enough to break through the surface tension film. The suction corresponding to the air pressure breaking through the surface tension film in the maximum pore is the air-entry value. Residual water content exists due to the strong absorption effect of soil particles on water. This content has been introduced in the previous section about water in soils.

As shown in Figure 1.15, typical soil-water characteristic curves have two apparent inflection points, one of which corresponds to air-entry value and the other of which corresponds to residual water content. Air-entry value and residual water content can be determined based on soil-water characteristic curves. The general procedure is: (1) draw a tangent line at the high water content section of the soil-water characteristic curve; (2) draw a vertical line at the saturated water content that is parallel to the axial of "soil suction"; and (3) take the suction value corresponding to the cross point of those two lines as the air-entry value. Similarly, a tangent line of the soil-water characteristic curve in low suction range is drawn at the inflection point corresponded to the high suction section; the curve in the high suction range is approximated to a straight line; and the water content corresponding to the cross point of these two lines is used as the residual water content. For all types of soils, the total suction of the soil with zero water content is the same. Via the experiments on various types of soils, Croney and Coleman (1961) measured a unique suction value slightly lower than 10^6 kPa. Vanapalli and Fredlund (Vanapalli *et al.*, 1996) measured soil-water characteristic curves with pressure plate tests and further verified the existence of this suction. Richard (1966) supported the selection of this soil suction with thermodynamic analysis.

Possessing the inflection points of air-entry value and residual water content is a characteristic of SWCC. Air-entry value and residual water content are two key characteristic points of SWCC. The curve between these two points presents smooth variation, of which the trend remains constant.

Another characteristic of SWCC is related to the process, i.e., whether it is desorption or absorption. In constant temperatures, the process that saturated soil samples gradually lose water until they only contain residual water content is the desorption process, and the measured SWCC is called the desorption curve; while the opposite process is the absorption process, and the measured SWCC is called the absorption curve. The absorption curve is under the desorption curve, which means that the degree of saturation of desorbing soils is higher than that of absorbing soil at the same suction level. When the suction of soils reduces to 0,

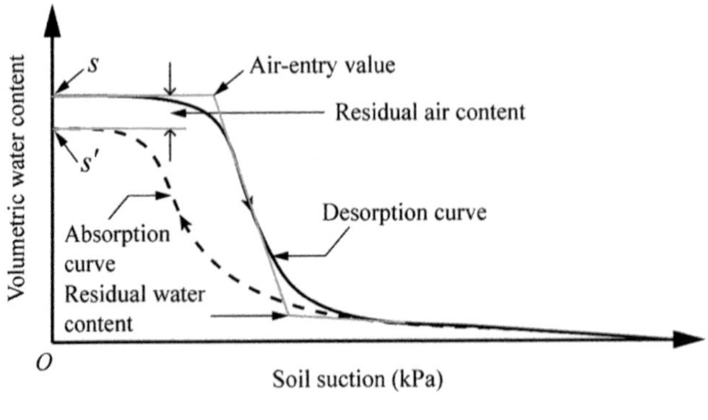

Figure 1.15 Typical soil-water characteristic curve of silty soils

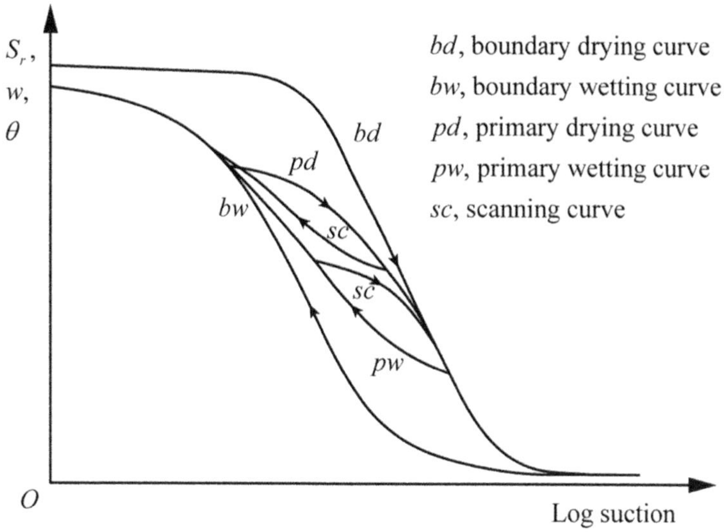

Figure 1.16 Hysteresis phenomena of soil-water characteristic curves

there is still a portion of pore air existing in pore water in terms of air bubbles. The degree of saturation of the soil is lower than 1.0 at this time. This phenomenon is called hysteresis of SWCC, as illustrated in Figure 1.16. The causes for the hysteresis phenomena are very complicated. The current three theories elucidating this phenomenon include bottle neck (ink bottle) theory, contact angle theory and delayed formation theory of meniscus (Lei *et al.*, 1988). As presented in Figure 1.16, the curves between the desorption curve and absorption curve are called scanning curves. When soils start to desorb water or reabsorb water from the partially moisturized status, the correlation curve between suction and water content varies following the scanning curves. It can be observed that the relationship between the suction of water in soils and water content is not a single value function. Furthermore, this relationship presents complex variation depending on desorption and absorption history.

1.6.2 Influence factors of SWCC

There are many factors influencing SWCC, mainly including mineral composition and particle constitution of soils, size of soil pores, and pore distribution, temperature, and stress status and stress history of soils.

1. The effect of mineral composition and particle constitution of soils on SWCC. Generally, corresponding to the same water content, the stronger the water absorption of the soil minerals is, the higher the suction is, while the higher the content of fine particles in soils is, the higher the water content of soils at the same level of suction, as shown in Figure 1.14.
2. The effect of the density and void ratio of soils on SWCC. For the same soil, its density and void ratio determine its pore size and distribution, which has a determining influence on the soil-water characteristic curves. Figure 1.17 presents the SWCCs of the same soil with different dry unit weights. The more compact the soil is, the lower the number of

large pores, while the higher the number of small pores. Therefore, at the same level of suction, the soils with higher dry unit weights generally contain higher water content. The density and pore ratio of soils show a particularly apparent effect on SWCC in low suction range.

3. The effect of temperature on SWCC. When the temperature increases, the viscosity and surface tension of water decrease and the matric potential correspondingly increases, or the suction in soils decreases. This effect is more apparent at low water content.

Besides the abovementioned factors, the stress status and stress history of soils also affect SWCC. Figure 1.18 shows the SWCC at different stress status.

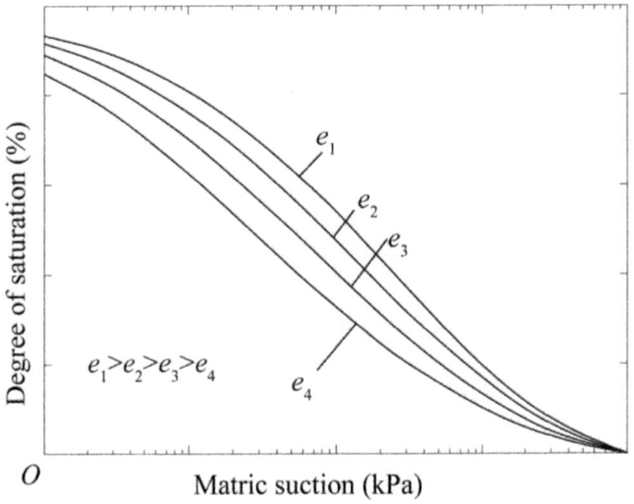

Figure 1.17 Soil-water characteristic curves of the same soil with different void ratios

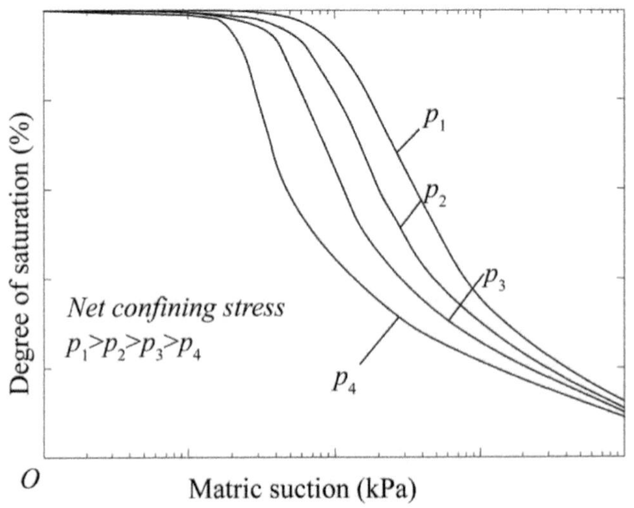

Figure 1.18 Soil-water characteristic curves at different stress status

1.6.3 SWCC and water content distribution in horizontal soil layers

SWCC can be determined via soil-water characteristic tests. They vary with desorption process or absorption process. They reflect the relationship between water content (degree of saturation) and matric potential. For a determined absorption process or desorption process, as long as the water content is known, the corresponding matric potential in a static equilibrium condition can be obtained according to SWCC. On the other hand, if the matric potential is known, the corresponding water content in a stable state can be determined. SWCC reflects the native water retaining property of soils, which is another constitutive relationship of soils.

The homogeneous soil layer in Figure 1.19 is used as an example to discuss the relationship between pore water pressure distribution (the pressure potential of saturated soils and matric potential of unsaturated soils) and water content distribution in the condition of constant temperature and hydrostatic equilibrium.

Constant temperatures mean no heat transfer, and thus the change of temperature potential can be neglected; hydrostatic equilibrium means that it is in static equilibrium, while the soil-water potentials at any two points in the soil layer are the same, i.e., there is no potential energy difference. The free water level in the soil layer is used as the standard reference plane. The state of pure free water on this plane is the standard reference state. Therefore, the soil-water potential of point O is zero. Point A below the underground water level in saturated soils is randomly taken, of which the vertical coordinate is z. Since the soil-water potential of point A is the same as point O, there is no potential energy difference between A and O. The soil-water potential of point A is:

$$\psi_{\mathrm{p}} + \psi_{\mathrm{g}} = 0 \tag{1.64}$$

Consequently:

$$\psi_{\mathrm{p}} = -\psi_{\mathrm{g}} = -\rho_{\mathrm{w}}\, g z \tag{1.65}$$

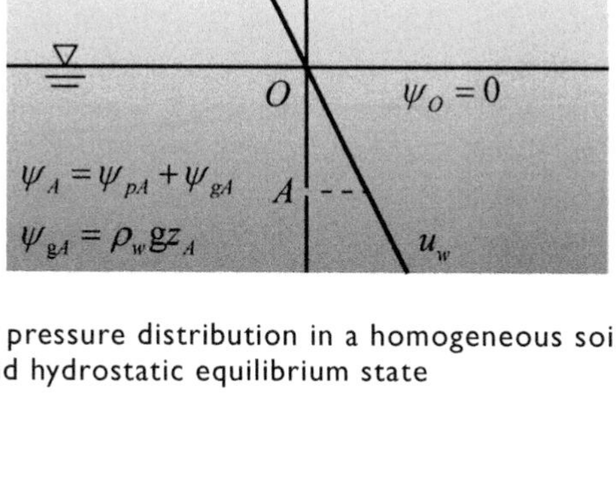

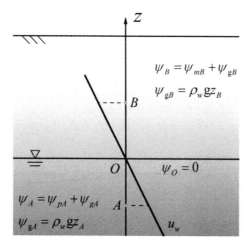

Figure 1.19 Pore water pressure distribution in a homogeneous soil layer in constant temperature and hydrostatic equilibrium state

When it is expressed with unit volume of water:

$$u_w = \psi_p = -\rho_w g z \tag{1.66}$$

Point B is randomly taken above the underground water level in unsaturated soils, of which the vertical coordinate is z. In open air, the pressure potential (air pressure potential) ψ_p is zero. Without counting solute potential, based on the fact that the soil-water potential of point B is equal to that of O:

$$\psi_m + \psi_g = 0 \tag{1.67}$$

and thus:

$$\psi_m = -\psi_g = -\rho_w g z \tag{1.68}$$

When unit volume of water is used to express the potential energy of water:

$$u_w = \psi_m = -\rho_w g z \tag{1.69}$$

This illustrated that matric potential presents linear distribution along the height z in the hydrostatic equilibrium state for unsaturated soil layers.

As a condition for static pore water (in static equilibrium), the equivalency between the soil-water potentials of two points in soil layers is meaningful only when pore water is connected, i.e., pore water can transfer pressure. As aforementioned, in unsaturated soils, when the water content is higher than the residual water content, pore water can transfer pressure, which is considered connected. When pore water is connected, its static equilibrium requires that the soil-water potentials of any two points in the soil layer are equal to each other. If there is soil-water potential difference between two points, the pore water will move. In other words, the static state of pore water indicates that the two points have no soil-water potential difference. According to the previous discussion, in homogeneous soil layers, when pore water maintains static equilibrium, matric potential is equal to negative gravity potential. Therefore, the water content of the soil layer distributes along the height z above the underground water level. This means that the relationship between water content and height z can represent the relationship between water content and matric potential. However, as we know, the relationship between the water content (degree of saturation) of soils and matric potential (matric suction) in static equilibrium is called the soil-water characteristic curve. Therefore, for the soil layer in which pore water is in a static equilibrium state, the distribution of water content along the height follows the SWCC, as shown in Figure 1.20. In other words, once the SWCC is obtained, the water content distribution along the height in the hydrostatic equilibrium state can be predicted. This distribution can be different with the variation of desorption or absorption process that the soil layer experiences. However, it is fixed. This means that the SWCC can be obtained by measuring the steady water content distribution of a soil column above the hydrostatic level in constant temperatures.

It can be ascertained that under the conditions of constant temperatures and hydrostatic equilibrium, once the water content of the soils (e.g., soil samples) is known, the matric potential can be determined. In a case where the matric potential of the soils is known, the location of the soil samples in actual soil layers can be deduced, which is its height above the

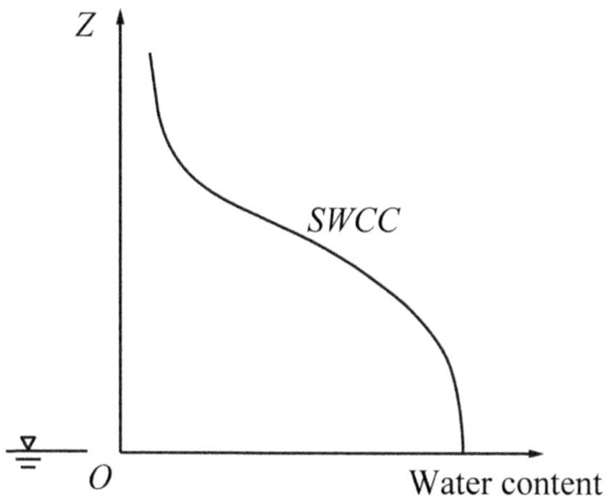

Figure 1.20 Distribution of water content along the height (the pore water is placed in static equilibrium soil layers)

underground water level. Regardless of the current location of the soil samples, their heights in actual soil layers are the same. On the other hand, under the conditions of constant temperatures and hydrostatic equilibrium, the soil layers at a certain height correspond to a fixed matric suction, and thus fixed water content.

1.7 Main contents of this book

This book has five chapters. The first chapter is an introduction, including two parts. In the first part, it briefly introduced the research achievements and existing problems about the effective stress of saturated and unsaturated soils and the equilibrium differential equations of soils used or proposed by some researchers. The second part is preparative knowledge. It mainly discussed the definition and continuity of the physical mechanical quantities of soils and introduced continuous medium matter models of soils and the definition of the soil skeleton area and pore area at a sectional area. In the section about the constitution of soils, we proposed that the pore water that is tightly combined with soil skeleton particles to bear and transfer loadings should be treated as a constituent of the soil skeleton, so-called soil skeleton water. In that section, the definitions of soil-water potential, its components and knowledge about the soil-water characteristic curves (SWCC) were also introduced. The section elucidated that the water content distribution curve along heights in the soil layer in semi-infinite space above the underground water level is SWCC.

The second chapter presents the equilibrium differential equation of soils expressed with total stress, and the equilibrium differential equations of the soil skeleton, pore water and pore air, including that of nearly saturated soils containing air bubbles. The key of this chapter is the force analysis on the free body of the soil skeleton. In the internal force analysis, we separately considered the pore fluid pressure and the stress induced by the other external forces.

In Chapter 3, the effective stress equations of saturated and unsaturated soils are first derived with equilibrium differential equations. It indicates the physical meaning of effective stress and discusses the formula of the volumetric strain and shear strength of soils. Subsequently, it verifies that the effective stress of unsaturated soils determines its shear strength with the shear strength testing results of unsaturated soils published in some literature. This chapter also discusses the effective stress principle for unsaturated soils and the simple application of the effective stress equation of unsaturated soils in the calculation of soil layer stress.

Chapter 4 shows the derivation of seepage equation of soils. It emphasizes how the equation of Darcy's law was derived from the equilibrium differential equation of pore water. It first introduces the derivation of Darcy's law formula for the seepage in saturated soils and the seepage equation and then the derivation of the seepage equation for unsaturated soils and the coefficient of permeability. The third section in this chapter discusses and modifies the equation of seepage force in current soil mechanics textbooks, while the fourth section provides the overflow conditions for gas in soils.

Chapter 5 discusses a number of questions that are related to effective stress, which are important for soil mechanics and still controversial, including (1) whether Terzaghi's effective stress equation needs to be modified; (2) whether effective stress is real stress or pseudo stress; (3) the relationship between effective stress and stress state variables; (4) the relationship between effective stress and soil skeleton stress; (5) the effective stress of unsaturated soils; and (6) whether the contractile skin of unsaturated soils should be treated as the fourth phase.

Chapter 2

Equilibrium differential equations of a soil

Equilibrium differential equations (abbreviated as equilibrium equations) of a soil, serving as the foundation for the theoretical system of soil mechanics, are the basis to study stress, strength and deformation of the soil. These equations include the equilibrium equation of soil mass and each phase of the soil, i.e., the soil skeleton, pore water and pore air. The equilibrium equation of soil mass represents the momentum (stress) equilibrium conditions on the free body of soil mixture expressed by total stress; the equilibrium equation of the soil skeleton represents the force equilibrium conditions on the free body of the soil skeleton expressed by skeleton stress; and the saturation has a similar relationship to pore water and pore air.

In porous medium mechanics and mixture theory, soil skeleton stress is defined as the internal force of the soil skeleton on a unit area of soil mass, i.e., the average of the total internal force of the soil skeleton on a unit area of soil mass; or it could be defined as the internal force on a unit area of the soil skeleton, i.e., the average of the total internal force of the soil skeleton on a unit area of the soil skeleton. Both of them are the average values of all the external forces acting on the soil skeleton. Their difference is only whether this force is averaged by the area of soil mass or of the soil skeleton. Later in this book we will prove that the effective stress is the internal force of the soil skeleton on a unit area of soil mass induced by all the external forces, excluding pore fluid pressure, i.e., the skeleton stress which does not include that due to pore fluid pressure.

2.1 Equilibrium differential equations of soil mass

As a soil mass, the soil mixture is usually considered to include the soil skeleton and pore fluid. The stress of soil mass is called the total stress of the soil. The total stress equilibrium equations of soil mass can be obtained by equilibrium analysis of the internal forces on the free body of soil mass.

While taking the free body of soil mass for stress analysis, the internal forces (stresses) exposed on the surfaces of the free body are the total stress, as shown in Figure 2.1, in which only the stresses on the xOz plane of the free body are presented for simplicity (Shao, 2011a, 2012).

The equilibrium differential equations for the free body of soil mass in total stress can be obtained based on the equilibrium conditions, as follows:

$$\begin{cases} \dfrac{\partial \sigma_{tx}}{\partial x} + \dfrac{\partial \tau_{zx}}{\partial z} + X_{swx} = 0 \\ \dfrac{\partial \tau_{xz}}{\partial x} + \dfrac{\partial \sigma_{tz}}{\partial z} + X_{swz} = 0 \end{cases} \tag{2.1}$$

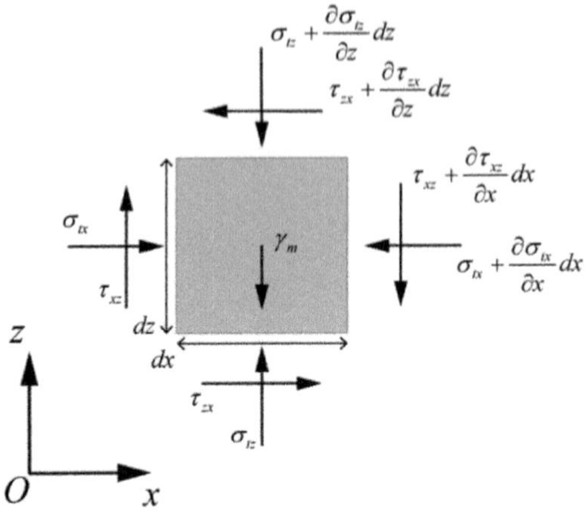

Figure 2.1 A free body of soil mass and the diagram of stress

Extending it to a three-dimensional case, the equilibrium differential equations of soil mass in total stress are:

$$\begin{cases} \dfrac{\partial \sigma_{tx}}{\partial x} + \dfrac{\partial \tau_{yx}}{\partial y} + \dfrac{\partial \tau_{zx}}{\partial z} + X_{swx} = 0 \\ \dfrac{\partial \tau_{xy}}{\partial x} + \dfrac{\partial \sigma_{ty}}{\partial y} + \dfrac{\partial \tau_{zy}}{\partial z} + X_{swy} = 0 \\ \dfrac{\partial \tau_{xz}}{\partial x} + \dfrac{\partial \tau_{yz}}{\partial y} + \dfrac{\partial \sigma_{tz}}{\partial z} + X_{swz} = 0 \end{cases} \tag{2.2}$$

where σ_{tx}, σ_{ty}, σ_{tz} are the total normal stresses, τ_{xy}, τ_{yz}, τ_{zx} are the shear stresses and X_{swx}, X_{swy}, X_{swz} are the body force components of soil mass. When there is only gravity acting on soil mass, $X_{swx} = X_{swy} = 0$; for saturated soils, $X_{swz} = \rho_{sat}$ g (ρ_{sat} is the saturated density of a soil, and g is the gravitational acceleration); for unsaturated soils, $X_{swz} = \rho$ g (ρ is the natural density of a soil).

2.2 Static equilibrium equation of the soil skeleton

The stress-strain problems could not be solved by using the equilibrium equation of total stress because of two reasons: (1) there is no definite correlation between the total stress and strain of a soil, and (2) the deformation and strength of a soil do not exclusively depend on the total stress. Therefore, to investigate the deformation and strength of soil mass, i.e., the deformation and strength of the soil skeleton, stress analysis on the soil skeleton is necessary.

2.2.1 *Force analysis on the free body of the soil skeleton*

Firstly, the forces on the free body of saturated soil skeleton are analyzed. Figure 2.2 (a) and (b) schematically show the equilibrium analysis of internal forces on the free body of the

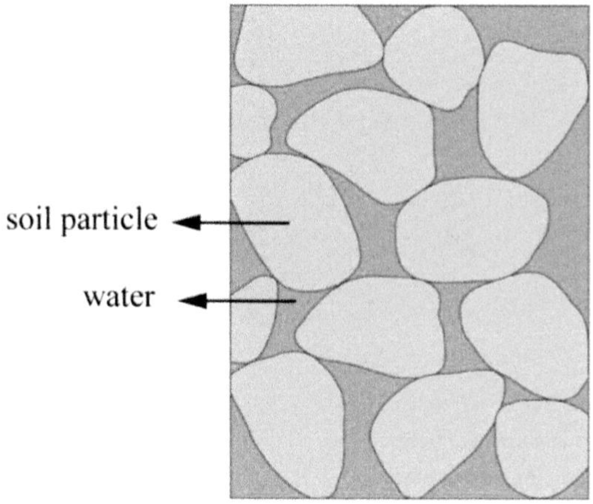

(a) The mixture of saturated soil mass

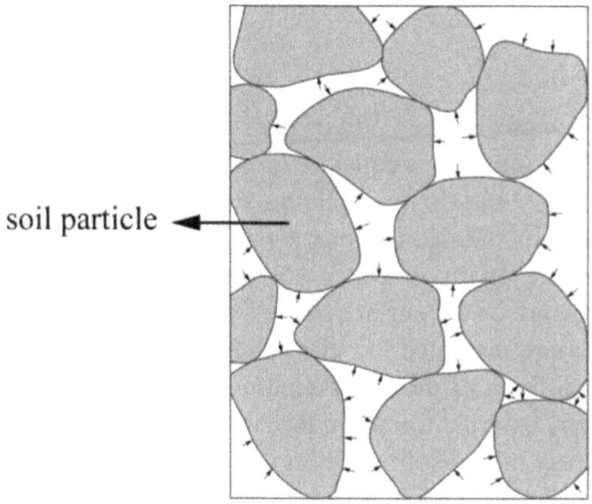

(b) Sketch of the effect of pore water pressure on the soil skeleton

Figure 2.2 The pore water pressure on the particles in the free body of the soil skeleton

soil skeleton taken from saturated soil mass. The pore water has to be removed from the soil skeleton when force analysis is conducted on the soil skeleton independently. In this case, the forces on soil skeleton due to pore water are exposed, including pore water pressure and the force on the soil skeleton induced by the flowing of pore water, i.e., seepage force. The former results in surface forces and acts on the surface of soil particles while the later results in interaction forces between the soil skeleton and pore water, as shown in Figure 2.2 (a) and (b). At first only the soil skeleton stress due to uniform pore water pressure is discussed here, which is called pore pressure-induced soil skeleton stress in the following text.

Figure 2.3 presents the internal force on the surface of the free body of the soil skeleton induced by uniform pore water pressure. As we can see, when uniform pore water pressure alone acts on soil particles, the average intensity of the exposed stress on any section of the particle and the stress on the interface between particles are both equal to pore water pressure.

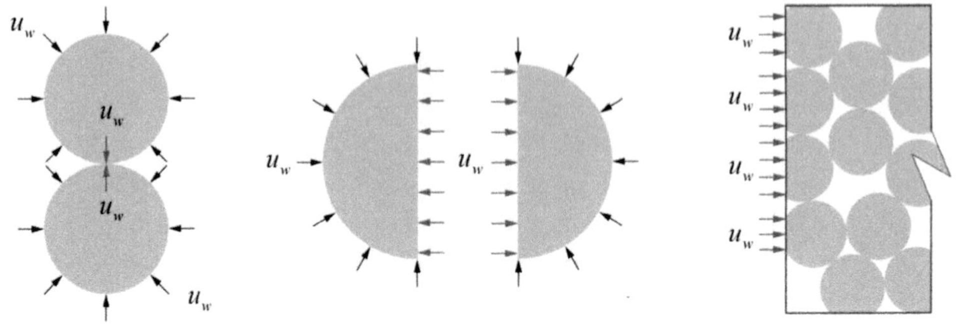

(a) Stress on the interface of the particles
(b) Stress on a section of the soil particle
(c) Stress on the section of the soil skeleton

Figure 2.3 The stresses on the soil particle and the soil skeleton induced by pore water pressure

Therefore, on the surface of the free body of the soil skeleton, the internal force of the soil skeleton due to pore water pressure is equal to the product of pore water pressure and the area of the soil skeleton, as $u_w(1 - n)A$. In this equation, u_w is the pore water pressure, n is the porosity, A is the sectional area of soil mass, and $(1 - n)A$ represents the area of the soil skeleton.

This analysis illustrates that pore water would result in exposed stress with an identical value of pore water pressure on the surface of the free body of the soil skeleton when the soil skeleton and pore water are separated for stress analysis individually.

Besides the internal force of the soil skeleton due to pore water pressure, there is also another stress induced by external forces on the surface of the free body of the soil skeleton, which is called the stress of the soil skeleton due to external forces. Let N and T denote the normal and tangential internal forces on a section induced by external forces excluding pore water pressure; then the soil skeleton stress due to external forces can be expressed by $\sigma = N/A$, $\tau = T/A$, where A is the total area of the section. So the soil skeleton stress due to external forces herein is the average internal force of the soil skeleton on the area of soil mass. This internal force is induced by all the external forces on the soil skeleton, excluding pore water pressure. It should be noted that taking the area of soil mass rather than that of the soil skeleton itself is mainly to facilitate the equilibrium analysis.

Therefore, the forces on the free body of the soil skeleton can be sketched as in Figure 2.4. For simplicity, only the forces on one section of the soil skeleton are shown. Based on the definition, the action area of soil skeleton stress due to external forces is the total area of the free body, while the action area of pore water pressure is the area occupied by the soil skeleton.

According to the analysis above, the forces acting on the soil skeleton can be classified into two categories: the pore water pressure and internal force of the soil skeleton induced by it, and external forces and the internal forces of the soil skeleton induced by them. Although both categories of forces can keep the soil skeleton in a state of equilibrium, their effects are different. The force of the first category has the properties of hydrostatic pressure, which only induces volumetric deformation of a soil particle. Furthermore, among these forces from the first category, only the stresses in interface between particles contribute to the shear strength

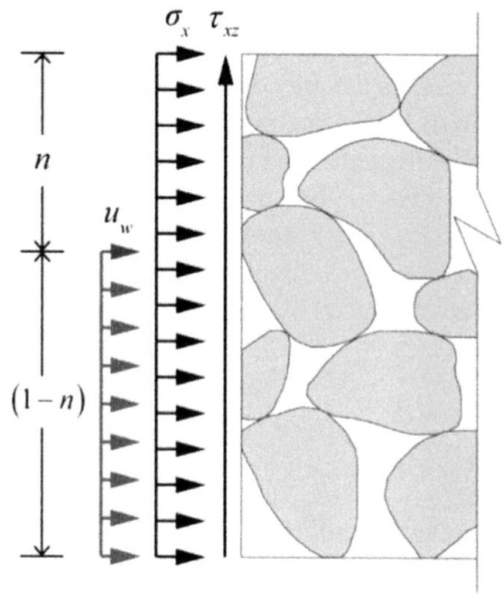

Figure 2.4 Force diagram of the surface of the free body of the soil skeleton

of the soil. When the pore water pressure is not significant, the volumetric change of the soil particle induced by pore water pressure can be neglected due to the great bulk modulus of soil particles. Meanwhile, the contribution of contact stress between soil particles on the strength of soil mass is also negligible since the contact area between particles is very small, in general. Consequently, the strength and deformation of the soil mass are exclusively governed by soil skeleton stress due to external forces.

Next, the forces on the free body of unsaturated soil skeleton are analyzed. When a free body of the soil skeleton is taken from the soil mass for internal force analysis independently, the effects of pore water and pore air on the soil skeleton will be revealed. These effects include the effects of pore water pressure and pore air pressure, the surface tension at the interface among water, air and soil particles, and the seepage force induced by pore water seepage and pore air seepage, respectively.

When it comes to the soil skeleton stresses induced by uniform pore water pressure and pore air pressure, the uniform pore fluid pressures possess normal stresses, without shear stresses, on the free body of the soil skeleton. If the quantitative value of the induced stresses is denoted as pore water pressure u_w and pore air pressure u_a, respectively, and the corresponding action areas on the soil skeleton are represented by A_{sw} and A_{sa}, then:

$$A_{sw} = (n_w/n)A_s \tag{2.3}$$

$$A_{sa} = (n_a/n)A_s \tag{2.4}$$

and:

$$A_s = (1-n)A \tag{2.5}$$

$$A_{sw} + A_{sa} = A_s \tag{2.6}$$

where A is the total area of the free body, A_s is the area of the soil skeleton, n is the porosity, n_w and n_a are the porosities corresponding to pore water and pore air, respectively; n_w/n and n_a/n are the fractions of pore water and pore air on the area occupied by pore fluid, respectively. The sum of these two action areas is the area of the soil skeleton.

According to the concept of matric suction, the effect of pore fluid pressure on the free body of the soil skeleton can be treated as the pore air pressure acts on the area of the entire soil skeleton, A_s; while the negative matrix suction $-(u_a - u_w)$ acts on the soil skeleton area occupied by pore water A_{sw}, as shown in Figure 2.5. For a better understanding of this standpoint, P is introduced to represent the resultant force of the internal forces on the surface of the free body of the soil skeleton induced by pore fluid pressure, as presented in Figures 2.6 and 2.7. Then P can be written as:

$$\begin{aligned} P &= u_a A_{sa} + u_w A_{sw} = u_a (A_s - A_{sw}) + u_w A_{sw} \\ &= u_a A_s + (u_w A_{sw} - u_a A_{sw}) = u_a A_s - (u_a - u_w) A_{sw} \end{aligned} \tag{2.7}$$

Similarly, the aforementioned pore fluid pressure can also be considered as that in which the pore water pressure acts on the area of the entire soil skeleton, A_s, and that in which the matric suction $(u_a\text{-}u_w)$ acts on the soil skeleton area occupied by pore air A_{sa}. The same procedure on P can be shown as follows:

$$\begin{aligned} P &= u_w A_{sw} + u_a A_{sa} = u_w (A_s - A_{sa}) + u_a A_{sa} \\ &= u_w A_s + (u_a A_{sa} - u_w A_{sa}) = u_w A_s + (u_a - u_w) A_{sa} \end{aligned} \tag{2.8}$$

The effect of surface tension is analyzed subsequently. The properties of surface tension are different from those of pore fluid pressure. Pore fluid pressure is a kind of surface force with the properties of hydrostatic pressure. It acts perpendicular to the surface of particles, and the magnitude is the same in every direction. However, the surface tension, which can be considered as a body force, belongs to the interaction force between the soil skeleton and pore water, as it is a concentrated force, just as gravity in the free body.

As shown in Figure 2.5, when water content is uniform, the resultant force of the surface tension on the free body of the soil skeleton is equal to 0, i.e., there is no additional body force. Otherwise, when water content is nonuniform, there is unbalanced surface tension acting on the free body of the soil skeleton, which can be combined with the seepage force of pore water as a body force when it comes to stress analysis.

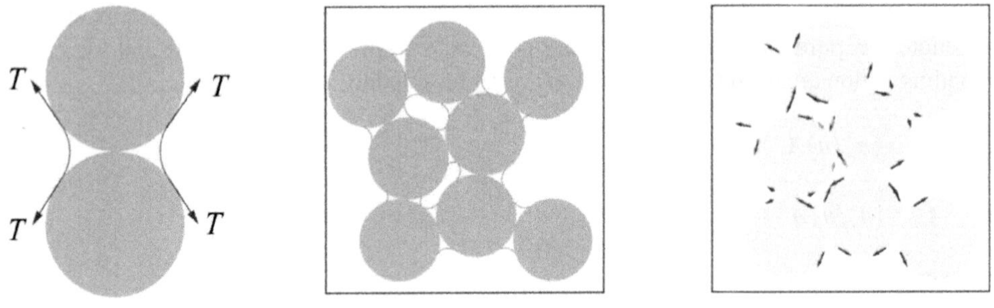

Figure 2.5 The effects of matric suction and surface tension

Finally, the seepage force is analyzed. Besides the effects of pore fluid pressure and surface tension, there is also the force induced by the relative motion between pore fluid and the soil skeleton, so-called seepage force, acting on the free body of the soil skeleton when pore fluid is moving. In the free body of the soil skeleton, the seepage force combined with unbalanced surface tension is denoted by $\vec{f}_{sw}$, and its components are f_{swx}, f_{swy} and f_{swz}, respectively.

In summary, Figure 2.6 and Figure 2.7 schematically illustrate the stresses on a vertical section of the free body of the unsaturated soil skeleton. Like saturated soils, the normal stress of the soil skeleton, σ_x, is defined as the internal force of the soil skeleton induced by the external forces on the unit area of soil mass, excluding pore fluid pressure. This normal stress is the soil skeleton stress due to external forces. As previously mentioned, Figure 2.7 shows the stress due to pore air pressure acting on the total area of the soil skeleton and the stress due to the matric suction acting on the area occupied by pore water. The resultant force of both these two stresses is equal to that of stresses induced by pore water pressure and pore air pressure, as shown in Figure 2.6.

2.2.2 Equilibrium differential equations of the independent phase of soil mass

Usually, soil mass is considered as a three-phase object, consisted of the soil skeleton, pore water and pore air. When completely saturated, it becomes a two-phase object constituted of the soil skeleton and pore water. The equilibrium differential equations of each independent phase of saturated soils and unsaturated soils are introduced in the following text.

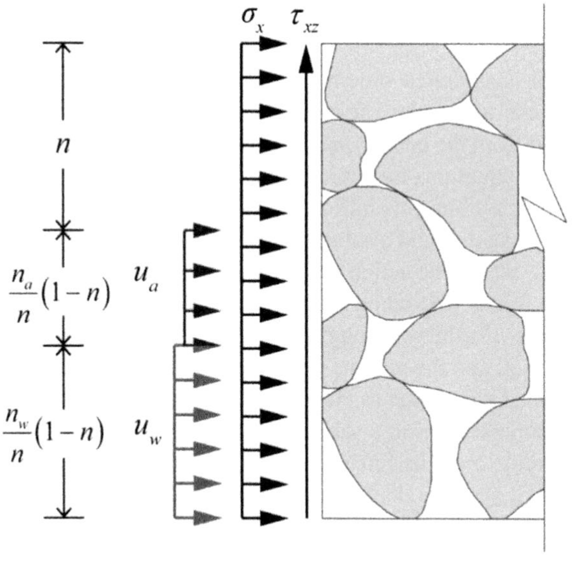

Figure 2.6 Sketch of the horizontal forces on the free body of the soil skeleton (the stresses due to pore water pressure and pore air pressure are applied on their respective occupied areas)

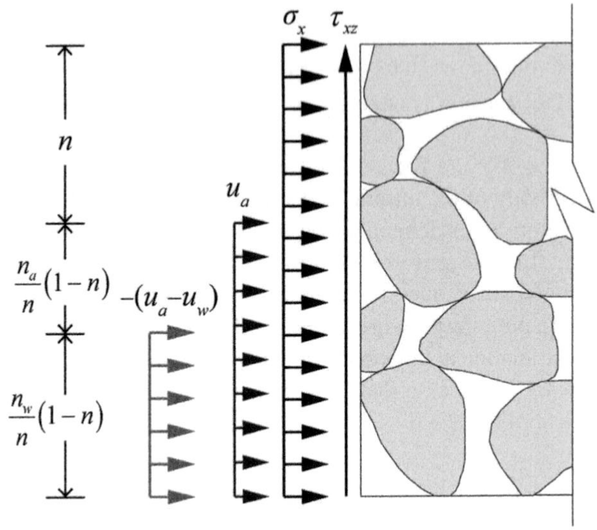

Figure 2.7 Sketch of the horizontal forces on the free body of the soil skeleton (the stress due to pore air pressure is applied on the area of the entire soil skeleton and the stress due to matric suction is applied on the area occupied by pore water)

2.2.2.1 Equilibrium equations for the soil skeleton and pore water of saturated soils

There is a saturated or unsaturated state for a soil. The saturated state may be regarded as a special case of the unsaturated state. The equilibrium differential equation for saturated states can be directly derived from the counterparts of unsaturated soils. For a better understanding, equilibrium differential equations for saturated soils are firstly introduced.

According to the stress analysis in Section 2.2.1, it is known that the internal forces of the free body of the soil skeleton in a saturated soil include (1) the soil skeleton stress due to pore water pressure, of which the action area is that of the soil skeleton; (2) the soil skeleton stress due to external forces (including normal stress and shear stress), of which the action area is that of soil mass; (3) the self-weight of the soil skeleton, equal to the product of the dry unit weight of the soil and the volume of soil mass; and (4) the interaction force between pore water and the soil skeleton due to the gradient of water potential. The force diagram of the free body of the soil skeleton in a saturated soil can be drawn accordingly, as shown in Figure 2.8. For simplicity, only the forces on the xOz plane of the free body are presented.

Figure 2.9 presents the force diagram of the free body of pore water in a saturated soil. The internal forces of the free body include (1) the pore water pressure on the free body's section that is perpendicular to the surface and acts on the area occupied by pore water; (2) the interaction force between pore water and the soil skeleton of the free body; (3) the self-weight of pore water, which is equal to the unit weight of water multiplied by the volume of pore water.

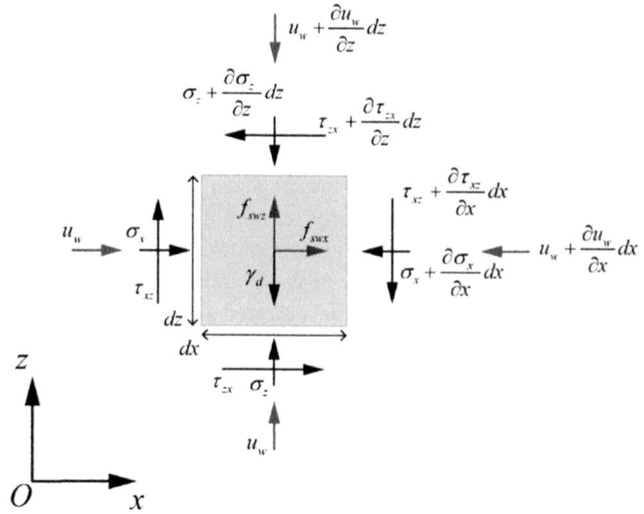

Figure 2.8 Stress analysis on the free body of the soil skeleton in a saturated soil

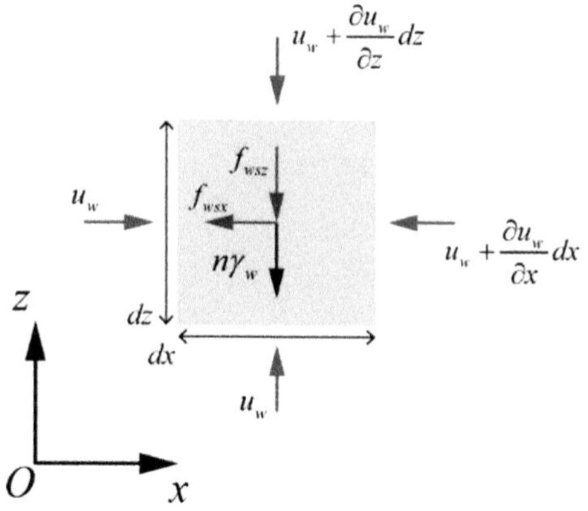

Figure 2.9 Stress analysis on the free body of pore water in a saturated soil

As illustrated in Figures 2.8 and 2.9, based on the equilibrium conditions for the free body, the equilibrium differential equations of the soil skeleton and pore water can be obtained, as:

$$\begin{cases} \dfrac{\partial \sigma_x}{\partial x} + \dfrac{\partial \tau_{zx}}{\partial z} + \dfrac{\partial \left[(1-n)u_w\right]}{\partial x} - f_{swx} = 0 \\ \dfrac{\partial \sigma_z}{\partial z} + \dfrac{\partial \tau_{xz}}{\partial x} + \dfrac{\partial \left[(1-n)u_w\right]}{\partial z} - f_{swz} + \rho_d \, g = 0 \end{cases} \tag{2.9}$$

$$\begin{cases} \dfrac{\partial(nu_w)}{\partial x} + f_{wsx} = 0 \\ \dfrac{\partial(nu_w)}{\partial z} + f_{wsz} + n\rho_w \text{g} = 0 \end{cases} \tag{2.10}$$

where n is the porosity, u_w is pore water pressure, σ_x and σ_z are normal stresses of the soil skeleton due to external forces and τ_{xz} and τ_{zx} are the shear stresses of the soil skeleton due to external forces; f_{swx} and f_{wsx} are the seepage force on the soil skeleton and its reaction force on pore water in x direction, and f_{swz} and f_{wsz} are the counterparts of f_{swx} and f_{wsx} in z direction; ρ_d is the dry density of the soil, ρ_w is the density of water and g is the gravitational acceleration.

The equilibrium differential equations for saturated soils without any interaction forces between phases can be obtained by adding Equation (2.9) and (2.10), i.e.:

$$\begin{cases} \dfrac{\partial\sigma_x}{\partial x} + \dfrac{\partial\tau_{zx}}{\partial z} + \dfrac{\partial u_w}{\partial x} = 0 \\ \dfrac{\partial\sigma_z}{\partial z} + \dfrac{\partial\tau_{xz}}{\partial x} + \dfrac{\partial u_w}{\partial z} + \rho_{sat}\text{g} = 0 \end{cases} \tag{2.11}$$

where ρ_{sat} is the saturated density of soil, and $\rho_{sat} = \rho_d + n\rho_w$.

Extending equation (2.11) to a three-dimensional case, it can be rewritten as:

$$\begin{cases} \dfrac{\partial\sigma_x}{\partial x} + \dfrac{\partial\tau_{yx}}{\partial y} + \dfrac{\partial\tau_{zx}}{\partial z} + \dfrac{\partial u_w}{\partial x} = 0 \\ \dfrac{\partial\tau_{xy}}{\partial x} + \dfrac{\partial\sigma_y}{\partial y} + \dfrac{\partial\tau_{zy}}{\partial z} + \dfrac{\partial u_w}{\partial y} = 0 \\ \dfrac{\partial\tau_{xz}}{\partial x} + \dfrac{\partial\tau_{yz}}{\partial y} + \dfrac{\partial\sigma_z}{\partial z} + \dfrac{\partial u_w}{\partial z} + \rho_{sat}\text{g} = 0 \end{cases} \tag{2.12}$$

Equation (2.12) is the equilibrium equation of saturated soil expressed by soil skeleton stresses due to external forces. It is the same with Biot's consolidation equation. However, Biot's consolidation equation was derived by introducing Terzaghi's effective stress equation into the equilibrium equations of soil mass in total stress.

2.2.2.2 Equilibrium differential equations of the soil skeleton of unsaturated soils

Based on the force analysis on the free body of the soil skeleton (see Section 2.2.1), it can be recognized that the forces on the free body of the soil skeleton in unsaturated soils include (1) pore fluid pressure-induced soil skeleton stress (only normal stress, perpendicular to the surface of the section of the soil skeleton). Its magnitudes are equal to pore water pressure and pore air pressure (the sum of the action area of these two pressures is the area of the soil skeleton); (2) the soil skeleton stress due to external forces (including normal stress and shear stress, of which the action area is the area of soil mass); (3) the self-weight of the soil skeleton (i.e., the product of the dry unit weight and the volume of soil mass); and (4) the interaction force between pore water and the soil skeleton due to the change of surface tension and seepage of pore water, as well as its counterparts induced by pore air.

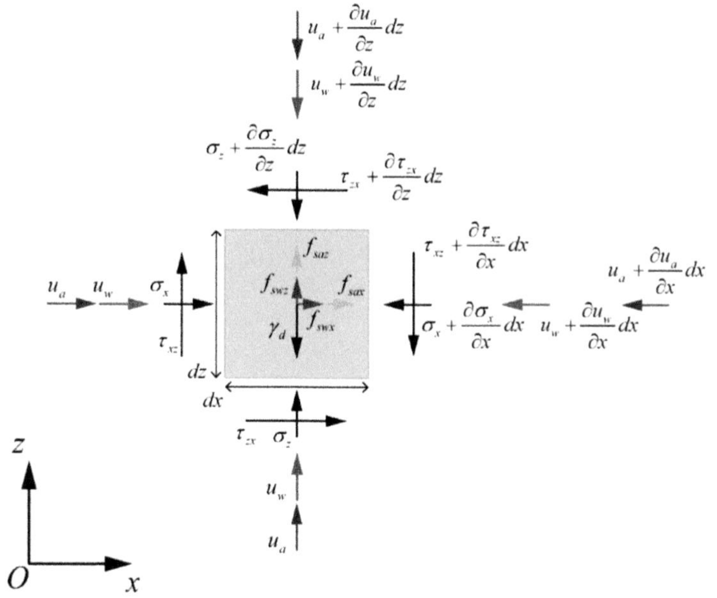

Figure 2.10 Stress and force on the free body of the unsaturated soil skeleton

Based on the equilibrium condition of the free body shown in Figure 2.10, the equilibrium differential equations of the soil skeleton can be obtained:

$$\begin{cases} \frac{\partial \sigma_x}{\partial x} + \frac{\partial \tau_{zx}}{\partial z} + \frac{1-n}{n}\frac{\partial (n_w u_w)}{\partial x} + \frac{1-n}{n}\frac{\partial (n_a u_a)}{\partial x} - f_{swx} - f_{sax} = 0 \\ \frac{\partial \tau_{xz}}{\partial x} + \frac{\partial \sigma_z}{\partial z} + \frac{1-n}{n}\frac{\partial (n_w u_w)}{\partial z} + \frac{1-n}{n}\frac{\partial (n_a u_a)}{\partial z} - f_{swz} - f_{saz} + \rho_d g = 0 \end{cases} \tag{2.13}$$

where f_{swx} is the acting force on the soil skeleton by pore water in x direction, f_{sax} is the seepage force of pore air on the soil skeleton in x direction, and f_{swz} and f_{sax} are the counterparts of f_{swx} and f_{sax} in z direction; n is the porosity of soil mass, and n_w and n_a are the porosity corresponding to pore water and pore air, respectively. The sum of these two porosities is n; σ_x and σ_z are the normal stresses of the soil skeleton due to external forces, and τ_{xz} and τ_{zx} are the shear stresses of the soil skeleton due to external forces. Similar to the definition of the stresses of saturated soils, these stresses are the soil skeleton stresses induced by the external forces, excluding pore fluid pressure. The dry density of the soil is ρ_d.

2.2.2.3 Equilibrium equations of pore water and pore air in unsaturated soils

The forces diagrams of the free body of pore water and pore air in unsaturated soils are shown in Figure 2.11. The forces applied on pore water/air include (1) the pore water/air pressure on the surface of the free body. It acts on the area occupied by pore water/air; (2) the interaction force between pore water/air and the soil skeleton; and (3) the self-weight of pore water/air, which is equal to the product of the unit weight of water/air and the volume of pore water/air.

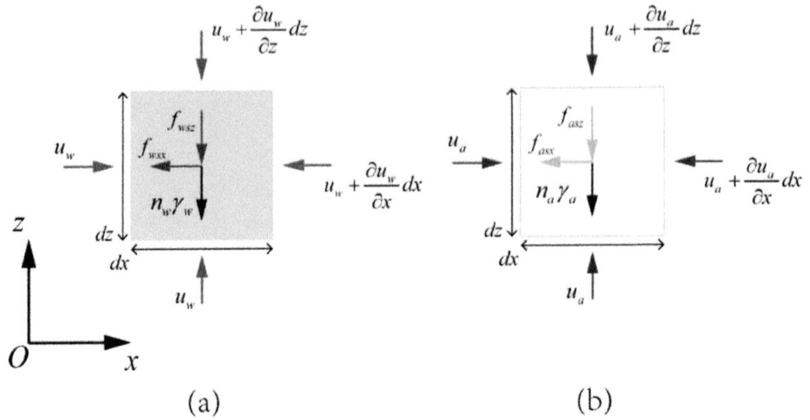

Figure 2.11 Stress and force on the free body of pore water (a) and pore air (b)

Based on equilibrium conditions of the free bodies shown in Figure 2.11, the equilibrium differential equations of the pore water and pore air in unsaturated soils are obtained (Equation (2.14) and (2.15), respectively):

$$\begin{cases} \dfrac{\partial(n_w u_w)}{\partial x} + f_{swx} = 0 \\ \dfrac{\partial(n_w u_w)}{\partial z} + f_{swz} + n_w \rho_w g = 0 \end{cases} \tag{2.14}$$

$$\begin{cases} \dfrac{\partial(n_a u_a)}{\partial x} + f_{sax} = 0 \\ \dfrac{\partial(n_a u_a)}{\partial z} + f_{saz} + n_a \rho_a g = 0 \end{cases} \tag{2.15}$$

where n_w and n_a are the porosity corresponding to pore water and pore air, respectively; u_w is the pore water pressure; u_a is the pore air pressure; f_{swx} and f_{swz} are the seepage force and unbalanced surface tension acting on the soil skeleton in x and z direction, respectively; f_{wsx} and f_{wsz} are the forces acting on pore water applied by the soil skeleton in x and z direction, with the same magnitude, but in opposite directions compared to f_{swx} and f_{swz}; f_{sax} and f_{saz} are the forces acting on the soil skeleton by the seepage of pore air in x and z direction; ρ_w is the density of pore water, and ρ_a is the density pore of air.

Adding Equations (2.13), (2.14) and (2.15) (equilibrium differential equation of the soil skeleton, pore water and pore air), the equilibrium differential equation of unsaturated soils without interaction forces between phases can be obtained:

$$\begin{cases} \dfrac{\partial \sigma_x}{\partial x} + \dfrac{\partial \tau_{zx}}{\partial z} + \dfrac{1}{n}\dfrac{\partial(n_w u_w)}{\partial x} + \dfrac{1}{n}\dfrac{\partial(n_a u_a)}{\partial x} = 0 \\ \dfrac{\partial \tau_{xz}}{\partial x} + \dfrac{\partial \sigma_z}{\partial z} + \dfrac{1}{n}\dfrac{\partial(n_w u_w)}{\partial z} + \dfrac{1}{n}\dfrac{\partial(n_a u_a)}{\partial z} + \rho g = 0 \end{cases} \tag{2.16}$$

where ρ is the natural density of the soil.

Since $S = n_w/n$ is the degree of saturation, Equation (2.16) can be rewritten as:

$$\begin{cases} \dfrac{\partial \sigma_x}{\partial x} + \dfrac{\partial \tau_{zx}}{\partial z} + \dfrac{\partial (S u_w)}{\partial x} + \dfrac{\partial ((1-S) u_a)}{\partial x} = 0 \\ \dfrac{\partial \tau_{xz}}{\partial x} + \dfrac{\partial \sigma_z}{\partial z} + \dfrac{\partial (S u_w)}{\partial z} + \dfrac{\partial ((1-S) u_a)}{\partial z} + \rho g = 0 \end{cases} \tag{2.17}$$

or:

$$\begin{cases} \dfrac{\partial \sigma_x}{\partial x} + \dfrac{\partial \tau_{zx}}{\partial z} + \dfrac{\partial u_a}{\partial x} - \dfrac{\partial (S(u_a - u_w))}{\partial x} = 0 \\ \dfrac{\partial \tau_{xz}}{\partial x} + \dfrac{\partial \sigma_z}{\partial z} + \dfrac{\partial u_a}{\partial z} - \dfrac{\partial (S(u_a - u_w))}{\partial z} + \rho g = 0 \end{cases} \tag{2.18}$$

Extending to a three-dimensional case, the equilibrium differential equation of unsaturated soils expressed by soil skeleton stress due to external forces can be obtained:

$$\begin{cases} \dfrac{\partial \sigma_x}{\partial x} + \dfrac{\partial \tau_{yx}}{\partial y} + \dfrac{\partial \tau_{zx}}{\partial z} + \dfrac{\partial (S u_w)}{\partial x} + \dfrac{\partial ((1-S) u_a)}{\partial x} = 0 \\ \dfrac{\partial \tau_{xy}}{\partial x} + \dfrac{\partial \sigma_y}{\partial y} + \dfrac{\partial \tau_{zy}}{\partial z} + \dfrac{\partial (S u_w)}{\partial y} + \dfrac{\partial ((1-S) u_a)}{\partial y} = 0 \\ \dfrac{\partial \tau_{xz}}{\partial x} + \dfrac{\partial \tau_{yz}}{\partial y} + \dfrac{\partial \sigma_z}{\partial z} + \dfrac{\partial (S u_w)}{\partial z} + \dfrac{\partial ((1-S) u_a)}{\partial z} + \rho g = 0 \end{cases} \tag{2.19}$$

or

$$\begin{cases} \dfrac{\partial \sigma_x}{\partial x} + \dfrac{\partial \tau_{yx}}{\partial y} + \dfrac{\partial \tau_{zx}}{\partial z} + \dfrac{\partial u_a}{\partial x} - \dfrac{\partial (S(u_a - u_w))}{\partial x} = 0 \\ \dfrac{\partial \tau_{xy}}{\partial x} + \dfrac{\partial \sigma_y}{\partial y} + \dfrac{\partial \tau_{zy}}{\partial z} + \dfrac{\partial u_a}{\partial y} - \dfrac{\partial (S(u_a - u_w))}{\partial y} = 0 \\ \dfrac{\partial \tau_{xz}}{\partial x} + \dfrac{\partial \tau_{yz}}{\partial y} + \dfrac{\partial \sigma_z}{\partial z} + \dfrac{\partial u_a}{\partial z} - \dfrac{\partial (S(u_a - u_w))}{\partial z} + \rho g = 0 \end{cases} \tag{2.20}$$

As mentioned in Chapter 1, a fraction of pore water in soils can withstand and transfer load together with soil particles. This fraction of pore water should be treated as a constituent of the soil skeleton. When this part of pore water is considered as a portion of the soil skeleton, the water content of soils refers to the effective water content, and the corresponding degree of saturation is the effective degree of saturation, which is expressed as follows:

$$S_e = \frac{S - S_r}{1 - S_r} \tag{2.21}$$

In this case, Equation (2.19) and (2.20) should be rewritten as:

$$\begin{cases} \frac{\partial \sigma_x}{\partial x}+\frac{\partial \tau_{yx}}{\partial y}+\frac{\partial \tau_{zx}}{\partial z}+\frac{\partial (S_e u_w)}{\partial x}+\frac{\partial ((1-S_e)u_a)}{\partial x}=0 \\ \frac{\partial \tau_{yx}}{\partial x}+\frac{\partial \sigma_y}{\partial y}+\frac{\partial \tau_{zy}}{\partial z}+\frac{\partial (S_e u_w)}{\partial y}+\frac{\partial ((1-S_e)u_a)}{\partial y}=0 \\ \frac{\partial \tau_{xz}}{\partial x}+\frac{\partial \tau_{yz}}{\partial y}+\frac{\partial \sigma_z}{\partial z}+\frac{\partial (S_e u_w)}{\partial z}+\frac{\partial ((1-S_e)u_a)}{\partial z}+\rho g=0 \end{cases} \tag{2.22}$$

or:

$$\begin{cases} \frac{\partial \sigma_x}{\partial x}+\frac{\partial \tau_{yx}}{\partial y}+\frac{\partial \tau_{zx}}{\partial z}+\frac{\partial u_a}{\partial x}-\frac{\partial (S_e (u_a-u_w))}{\partial x}=0 \\ \frac{\partial \tau_{xy}}{\partial x}+\frac{\partial \sigma_y}{\partial y}+\frac{\partial \tau_{zy}}{\partial z}+\frac{\partial u_a}{\partial y}-\frac{\partial (S_e (u_a-u_w))}{\partial y}=0 \\ \frac{\partial \tau_{xz}}{\partial x}+\frac{\partial \tau_{yz}}{\partial y}+\frac{\partial \sigma_z}{\partial z}+\frac{\partial u_a}{\partial z}-\frac{\partial (S_e (u_a-u_w))}{\partial z}+\rho g=0 \end{cases} \tag{2.23}$$

It should be noted that the equilibrium differential equation of the soil skeleton in saturated soils can be obtained when $S_e = 1$.

2.2.2.4 Equilibrium differential equations of nearly saturated soils containing air bubbles

In-situ saturated soils usually contain gas in the form of bubbles. If the size of a gas bubble exceeds that of Representative Elementary Volume (*REV*), the gas bubble actually becomes a closed boundary in the soil. It can no longer be considered as a part of the soil mass in this case. Certainly, the size of gas bubbles is generally small, and it can be designated as a micro gas bubble in soil mass. When the maximum diameter of the gas bubble is far smaller than the dimension of *REV* and uniformly distributed, soil mass can still be treated as a homogeneous medium.

Generally, in the nearly saturated soil with gas bubbles, a portion of these bubbles are adsorbed to the water membrane surrounding the soil particles, while the others are entirely surrounded by pore water. The former portion is called the adsorbed micro air bubbles, and the latter is called micro water air bubbles. The micro water air bubbles affect the density and compressibility of pore water, but not soil skeleton stresses, while the adsorbed air bubbles influence the stress and deformation of the soil skeleton. Now the equilibrium differential equations of the nearly saturated soils containing adsorbed micro air bubbles is derived by taking the free body of different phases for internal equilibrium analysis.

The force analysis on the free body of nearly saturated soil containing adsorbed micro air bubbles is shown in Figure 2.12. Besides pore water pressure, the adsorbed micro air bubbles exert pressure on the free body compared to that of the saturated soil skeleton. The other forces acting on the free body of these two types of soils are the same. Compared to the free body of the unsaturated soil skeleton, the free body in Figure 2.12 includes the effect of pore water pressure and pore air pressure, without the seepage force between pore air and the soil skeleton and the unbalanced surface tension. The sum of the action area of pore water pressure (i.e., the area occupied by pore water) and that of pore air pressure (i.e., the area

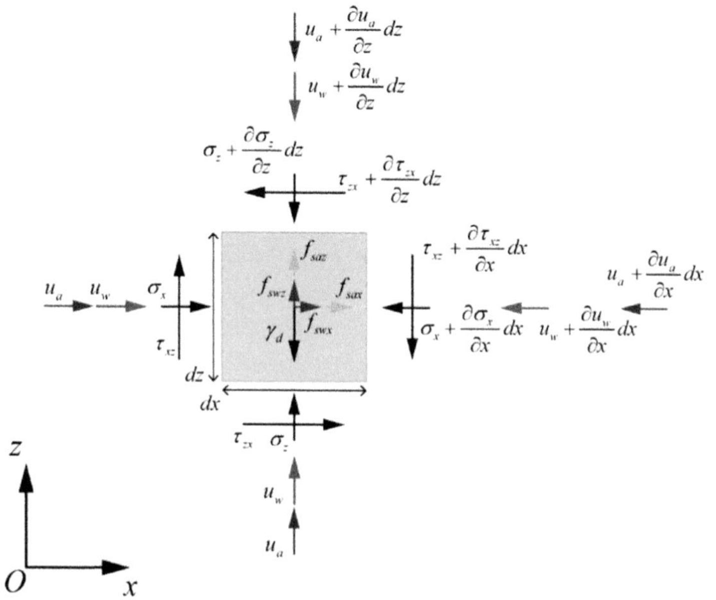

Figure 2.12 Force diagram of the free body of nearly saturated soils containing adsorptive micro gas bubbles

occupied by pore air) is the area of the soil skeleton. Based on the equilibrium conditions for the free body, the equilibrium differential equations of the soil skeleton are:

$$\begin{cases} \frac{\partial \sigma_x}{\partial x}+\frac{\partial \tau_{xy}}{\partial y}+\frac{\partial \tau_{xz}}{\partial z}+\frac{1-n_e}{n_e}\frac{\partial(n_{ew}u_w)}{\partial x}+\frac{1-n_e}{n_e}\frac{\partial(n_{ea}u_a)}{\partial x}-f_{swax}=0 \\ \frac{\partial \tau_{yx}}{\partial x}+\frac{\partial \sigma_y}{\partial y}+\frac{\partial \tau_{yz}}{\partial z}+\frac{1-n_e}{n_e}\frac{\partial(n_{ew}u_w)}{\partial y}+\frac{1-n_e}{n_e}\frac{\partial(n_{ea}u_a)}{\partial y}-f_{sway}=0 \\ \frac{\partial \tau_{zx}}{\partial x}+\frac{\partial \tau_{zy}}{\partial y}+\frac{\partial \sigma_z}{\partial z}+\frac{1-n_e}{n_e}\frac{\partial(n_{ew}u_w)}{\partial z}+\frac{1-n_e}{n_e}\frac{\partial(n_{ea}u_a)}{\partial z}-f_{swaz}+\rho_s g=0 \end{cases} \tag{2.24}$$

The micro bubbles adsorbed to the surfaces of soil particles or micro soil clusters are generally steady. Therefore, the micro bubbles can be combined with pore water to conduct internal equilibrium analysis. Assuming the adsorbed micro bubbles are uniformly distributed, the following equations can be obtained via the equilibrium analysis on the internal forces of the free body of pore water (as shown in Fig. 2.13):

$$\begin{cases} \frac{\partial(n_{ea}u_a)}{\partial x}+\frac{\partial(n_{ew}u_w)}{\partial x}+f_{swax}+X_{wax}=0 \\ \frac{\partial(n_{ea}u_a)}{\partial y}+\frac{\partial(n_{ew}u_w)}{\partial y}+f_{sway}+X_{way}=0 \\ \frac{\partial(n_{ea}u_a)}{\partial z}+\frac{\partial(n_{ew}u_w)}{\partial z}+f_{swaz}+X_{waz}=0 \end{cases} \tag{2.25}$$

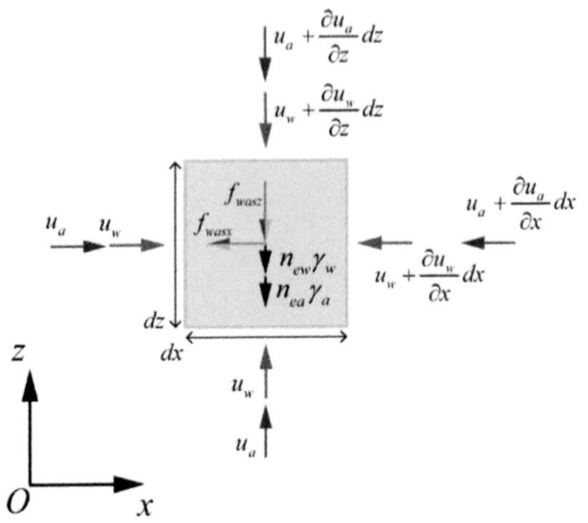

Figure 2.13 Force diagram of the free body of pore water in nearly saturated soils containing adhesive micro gas bubbles

where f_{swai} $(i = x, z)$ is the interaction force between the soil skeleton and the pore water and X_{wai} $(i = x, z)$ is the volumetric force on the unit volume of the pore water containing air bubbles.

When the effect of air content on the density of pore water is negligible, and taking into account only the self-weight of pore water, Equation (2.25) can be written as:

$$\begin{cases} \frac{\partial(n_{ea}u_a)}{\partial x} + \frac{\partial(n_{ew}u_w)}{\partial x} + f_{swax} = 0 \\ \frac{\partial(n_{ea}u_a)}{\partial y} + \frac{\partial(n_{ew}u_w)}{\partial y} + f_{sway} = 0 \\ \frac{\partial(n_{ea}u_a)}{\partial z} + \frac{\partial(n_{ew}u_w)}{\partial z} + f_{swaz} + n\rho_w g = 0 \end{cases} \tag{2.26}$$

Adding Equations (2.26) and (2.24), the following equations can be obtained:

$$\begin{cases} \frac{\partial\sigma_x}{\partial x} + \frac{\partial\tau_{xy}}{\partial y} + \frac{\partial\tau_{xz}}{\partial z} + \frac{\partial(S_e u_w)}{\partial x} + \frac{\partial((1-S_e)u_a)}{\partial x} = 0 \\ \frac{\partial\tau_{yx}}{\partial x} + \frac{\partial\sigma_y}{\partial y} + \frac{\partial\tau_{yz}}{\partial z} + \frac{\partial(S_e u_w)}{\partial y} + \frac{\partial((1-S_e)u_a)}{\partial y} = 0 \\ \frac{\partial\tau_{zx}}{\partial x} + \frac{\partial\tau_{zy}}{\partial y} + \frac{\partial\sigma_z}{\partial z} + \frac{\partial(S_e u_w)}{\partial z} + \frac{\partial((1-S_e)u_a)}{\partial z} + \rho g = 0 \end{cases} \tag{2.27}$$

or:

$$\begin{cases} \dfrac{\partial \sigma_x}{\partial x} + \dfrac{\partial \tau_{xy}}{\partial y} + \dfrac{\partial \tau_{xz}}{\partial z} + \dfrac{\partial u_a}{\partial x} - \dfrac{\partial \left(S_e \left(u_a - u_w\right)\right)}{\partial x} = 0 \\ \dfrac{\partial \tau_{yx}}{\partial x} + \dfrac{\partial \sigma_y}{\partial y} + \dfrac{\partial \tau_{yz}}{\partial z} + \dfrac{\partial u_a}{\partial y} - \dfrac{\partial \left(S_e \left(u_a - u_w\right)\right)}{\partial y} = 0 \\ \dfrac{\partial \tau_{zx}}{\partial x} + \dfrac{\partial \tau_{zy}}{\partial y} + \dfrac{\partial \sigma_z}{\partial z} + \dfrac{\partial u_a}{\partial z} - \dfrac{\partial \left(S_e \left(u_a - u_w\right)\right)}{\partial z} + \rho g = 0 \end{cases} \tag{2.28}$$

where S_e is the effective degree of saturation, and ρg is the overall unit weight of soil mass in z-direction.

It can be observed that Equation (2.28) is the same as the equilibrium equation of unsaturated soils, i.e., Equation (2.23).

Chapter 3

Effective stress

Terzaghi's effective stress equation is valid for saturated soil. A lot of works have been done for verifying the effective stress equation, and many different modified formulas of effective stress have been proposed up to now. This implies that people are not very clear about the physical meaning of effective stress.

For unsaturated soils, there are mainly two approaches for the study of effective stress: (1) The single effective stress approach. Measuring effective stress is critical for determining the strength and deformation of unsaturated soils, and the effective stress equation should be formulated similar to Bishop's form of the effective stress equation. (2) The stress state variables approach. Instead of single effective stress, certain independent stress state variables are proposed to determine the strength and deformation of unsaturated soils, such as net stress and matric suction. However, a new trend is to combine these two approaches, which is consistent with the idea that the strength and deformation of unsaturated soils are controlled by two stress state variables, including the effective stress component and the suction component, respectively.

The core of the research on unsaturated soils is the effective stress equation. Once the effective stress is determined, the theories of saturated soil mechanics established based on effective stress, such as strength, compressibility, earth pressure, constitutive relationship and stability analysis, can be directly introduced into unsaturated soils as the effective stress governs the strength and deformation of unsaturated soils.

The research work that had been done on effective stress can be classified into two aspects: first, to investigate the parameter χ in Bishop's effective stress equation for unsaturated soils, $\sigma' = \sigma - u_a + \chi (u_a - u_w)$, experimentally and theoretically to predict the shear strength and deformation (stiffness) of unsaturated soils (e.g., Alonso *et al.*, 2010; Khalili and Khabbaz, 1998; Khalili *et al.*, 2004); second, to derive the effective stress equation for unsaturated soils within the framework of thermodynamics (e.g., Lu *et al.*, 2010).

In this chapter, the definition and physical meaning of effective stress based on the equilibrium differential equation of soils (in Chapter 2) will be clarified. It indicates that effective stress is the soil skeleton stress induced by the external forces, excluding pore fluid pressure. Then, the correlation between effective stress and shear strength for unsaturated soils will be verified based on the test data published in the literature. A primary verification of the effective stress on the determination of deformation on unsaturated soils is also shown in the following texts. Finally, the simple application of effective stress on the stress calculation for the soil layers is presented in this chapter.

3.1 Effective stress equation and physical meaning of effective stress

3.1.1 Effective stress equation

Equation (2.2) is the equilibrium differential equation of soils in terms of total stress, and Equation (2.12) is the equilibrium differential equation of saturated soils expressed by the soil skeleton stress induced by external force, excluding pore fluid pressure. Comparison of these two equations leads to:

$$\sigma = \sigma_t - u_w \tag{3.1}$$

It should be noted that Equation (3.1) is exactly the same as Terzaghi's effective stress equation.

Equations (2.22) and (2.23) are the equilibrium equations of unsaturated soils expressed by soil skeleton stress, pore water pressure and pore air pressure. By comparing them to Equation (2.2), the relationship between the soil skeleton stress induced by external force, total stress, pore water pressure and pore air pressure can be formulated as:

$$\sigma = \sigma_t - S_e u_w - (1 - S_e) u_a \tag{3.2}$$

or:

$$\sigma = \sigma_t - u_a + S_e (u_a - u_w) \tag{3.3}$$

Where σ is the soil skeleton stress induced by external force, which excludes the pore fluid pressure, σ_t is the total stress, and u_a and u_w are pore air pressure and pore water pressure, respectively. $(u_a - u_w)$ is called matric suction and S_e is the effective degree of saturation.

Equations (3.2) and (3.3) are the effective stress equation that is applicable for both saturated soils and unsaturated soils. When the soil is saturated ($S_e = 1$), they degrade to Equation (3.1), i.e., Terzaghi's effective stress equation.

The effective stress equation reflects the relationship between soil skeleton stress induced by external force, total stress and pore fluid pressure, and this relationship is determined by the internal force (momentum) equilibrium condition. It is related to water content or degree of saturation of a soil but is irrelevant to the properties among soil particles, even to the interaction properties between soil particles and pore water.

It should be noted that the effective stress equation is identical to that proposed by Lu *et al.* (2010). However, Lu's effective stress is derived through the thermodynamic method. If the degree of saturation corresponding to the soil skeletal water content is the same as the definition of the micro degree of saturation (Alonso *et al.*, 2010), then the effective stress equation in Equation (3.3) is the same as the expression of the effective stress equation proposed by Alonso *et al.* (2010).

3.1.2 Physical meaning of effective stress

In the force analysis on the free body of the soil skeleton, the soil skeleton stress induced by pore fluid pressure and by the other external forces, i.e., pore pressure-induced soil skeleton stress and external force-induced soil skeleton stress, were considered separately. From the derivation of the effective stress equation, it can be clearly found that effective stress is the soil skeleton stress induced by external force, excluding pore fluid pressure (soil skeleton stress induced by external force, for short), which has explicit physical meaning.

As aforementioned, the pore pressure-induced soil skeleton stress and pore pressure leads only to the volumetric deformation of soil skeleton particles. In addition, for pore pressure, only the stress induced by it on the interfaces of the particles contributes to the shear strength of soils. In general, this kind of contribution to the strength and deformation of the soil skeleton can be neglected, and hence the shear strength and volume change of unsaturated soils is exclusively determined by the effective stress.

For saturated soils, there are two reasons to support the perspective that the soil skeleton stress induced by external force is Terzaghi's effective stress. First, it is formulated as the difference between the total stress and pore water pressure, with an identical expression of Terzaghi's effective stress. Second, when the effect of pore fluid pressure on the soil skeleton can be neglected, the strength and deformation of the soil skeleton are exclusively controlled by the soil skeleton stress induced by the external forces, while the effective stress principle states that "*the change in the volume and shear strength of soils is* exclusively *determined by the variation of effective stress*." Therefore, Terzaghi's effective stress is the soil skeleton stress induced by external forces.

For unsaturated soils, the contribution of pore pressure on the strength and deformation can be neglected in general. So the external forces-induced soil skeleton stress exclusively determines the strength and deformation. It is the effective stress.

On the other hand, what we can see here is that the effective stress equation is an identical relation about the total stress, effective stress and pore fluid pressure. The equilibrium differential equation of effective stress can be obtained directly with that of total stress by applying the effective stress equation. Furthermore, the effective stress at any point in the soil can be determined by the total stress, pore water pressure and pore air pressure.

3.2 Relationship between effective stress and shear strength/volumetric strain

3.2.1 *Formula of volumetric strain of a soil*

The deformation of a soil could precisely refer to that of the soil skeleton, as it is for the shear strength. Therefore, the strength and deformation are definitely related to the stress of the soil skeleton. As aforementioned, the soil skeleton stress can be divided into pore pressure-induced soil skeleton stress and external forces-induced soil skeleton stress. They have different contributions to the strength and deformation of a soil.

When it comes to soil's deformation, external forces-induced soil skeleton stress, i.e., effective stress, induces the deformation of the soil skeleton, which includes volumetric deformation and shear deformation, while pore pressure-induced soil skeleton stress only induces the volumetric deformation of soil particles. The total volumetric deformation consists of volumetric deformation caused by effective stress and soil particles' volumetric deformation caused by pore pressure. This can be illustrated by the following equation:

$$-\frac{\Delta V}{V}=C\{\Delta(\sigma_t-u_a)+\Delta[S_e(u_a-u_w)]\}+C_s\Delta u_a+C_sS_e\Delta(u_w-u_a) \tag{3.4}$$

The first term on the right side of the equation is the volumetric change induced by effective stress increment, while the second and third terms are the volumetric change of soil particles induced by the pore air pressure increment and matric suction increment, respectively. In Equation (3.4), C is the volumetric compression coefficient of soil (soil skeleton) and C_s is the volumetric compression coefficient of soil particle.

Equation (3.4) can be further rearranged to:

$$-\frac{\Delta V}{V}=C\cdot\left\{\Delta\sigma_t-\left(1-\frac{C_s}{C}\right)\Delta\left[u_a-S_e\left(u_a-u_w\right)\right]\right\} \tag{3.5}$$

This is the unified equation of volumetric strain for saturated and unsaturated soils. Further remarks on Equation (3.5) lead to: (1) the volumetric compression coefficient of soil particles is larger than that of the soil skeleton; (2) in practice, the pore air is usually connected to the atmosphere, so the pore air pressure can be neglected. Because of the negative correlation between suction and water content, when the matric suction is large, the effective degree of saturation would be relatively low, so the effect of matric suction on the compressibility of unsaturated soils can always be neglected due to the extreme value of the compression coefficient of soil particle.

Consequently, only for saturated soils, the effect of pore pressure on volumetric change needs to be considered when the pore water pressure is quite high. So, in general, the volumetric strain of a soil can be illustrated by:

$$-\frac{\Delta V}{V}=C\cdot\left[\Delta\sigma_t-\Delta u_a+S_e\left(u_a-u_w\right)\right] \tag{3.6}$$

It means that effective stress determines the volumetric change of a soil.

3.2.2 Shear strength formula of a soil

Like the volumetric behavior, the shear strength of a soil, i.e., the shear strength of the soil skeleton, also includes two parts, namely, soil skeleton strength corresponding to effective stress and the soil skeleton strength corresponding to the stress on the interfaces of soil particles, which is induced by pore fluid pressure. It can be expressed by the following formula:

$$\tau_f=c'+\left[(\sigma_t-u_a)+S_e\left(u_a-u_w\right)\right]\tan\varphi'+a_c u_a\tan\psi+a_c S_e\left(u_w-u_a\right)\tan\psi \tag{3.7}$$

where a_c is the ratio of contact area of skeleton particles to the whole area of soil, ψ is the internal friction angle for the contact interface of soil particle and S_e is the effective degree of saturation. The first term on the right side of equation (3.7) shows the contribution of cohesion, the second is the contribution of effective stress, the third is that of pore air pressure and the fourth is that of matric suction.

Equation (3.7) can be further rearranged to

$$\tau_f=c'+\left\{\sigma_t-(1-\frac{a_c\tan\psi}{\tan\varphi'})\left[u_a-S_e\left(u_a-u_w\right)\right]\right\}\tan\varphi' \tag{3.8}$$

This is a unified shear strength formula for saturated and unsaturated soils. Since the contact area among soil particles is extremely small in general, the contribution to the shear strength by pore fluid pressure can be neglected. In this case, the shear strength formula is:

$$\tau_f = c' + \left[\sigma_t - u_a + S_e\left(u_a - u_w\right)\right]\tan\varphi' \tag{3.9}$$

This illustrates that the statement "*the change in the volume and shear strength of soils is* exclusively *determined by the variation of effective stress*" is conditional. It is based on the premise that the effect of the pore pressure on the strength and deformation can be ignored.

Letting $S_e = 1$ in Equation (3.5) and (3.8), the equations for the volume change and shear strength of a saturated soil can be obtained as the effect of pore pressure on the strength and deformation is considered.

$$-\frac{\Delta V}{V} = C\left[\Delta\sigma_t - \left(1 - \frac{C_s}{C}\right)\Delta u_w\right] \tag{3.10}$$

$$\tau_f = c' + \left[\sigma_t - \left(1 - \frac{a_c \cdot \tan\psi}{\tan\varphi'}\right)u_w\right]\tan\varphi' \tag{3.11}$$

Equation (3.10) and (3.11) are the same as those proposed by Skempton (1961) with the purpose of indicating how Terzaghi's effective stress equation was derived and whether it needs to be modified. He concluded according to his experiment results that Terzaghi's effective stress equation is valid for saturated soil, whereas for saturated rock or concrete, it needs to be modified and can be formulated by Equations (3.10) and (3.11).

By introducing the concept of equivalent pore pressure proposed by Bishop *et al.*, and by means of the comparison of the shear strength equation for unsaturated soil (Equation 3.8) and its counterpart for saturated soil (Equation 3.11) and the comparison of the equations of volumetric change for unsaturated soil (Equation 3.5) and its counterpart for saturated soil (Equation 3.10), we can define an equivalent pore fluid pressure that can be expressed as:

$$u^* = u_a - S_e\left(u_a - u_w\right) \tag{3.12}$$

where u^* represents the equivalent pore fluid pressure.

Similar to the effective stress equation, the equivalent pore fluid pressure equation also has different forms; Equation (3.12) can be rearranged as:

$$u^* = \left(1 - S_e\right)u_a + S_e u_w \tag{3.13}$$

where the factors $(1 - S_e)$ and S_e are the corresponding effective degrees of saturation for pore air and pore water, respectively. Equation (3.13) can also be extended to the other kind of fluids in porous media, such as oil, air, water and other mutually insoluble fluids; it can be written as:

$$u^* = S_{ei}u_{wi}\left(i = 1, 2, \ldots, M\right) \tag{3.14}$$

where u^* is the equivalent pore fluid pressure, M is the total number of pore fluid phases, S_{ei} is the effective degree of saturation corresponding to the ith pore fluid phase, and u_i is the pressure of the ith pore fluid phase. The tensor notation here in the equation denotes the Einstein summation convention for the pore fluid pressure and the corresponding effective degree of saturation of each phase.

Based on the previous discussion, the effective stress equation for heterogeneous porous media can be expressed as:

$$\sigma = \sigma_t - u^* \tag{3.15}$$

It is applicable for a variety of porous media materials including saturated and unsaturated soils.

3.2.3 Effective stress and the constitutive relationship of a soil

The constitutive relationship of soils plays a key role in soil mechanics, especially in characterizing the stress and strain relationship of a soil. The deformation of a soil refers to that of the soil skeleton, and there is no specific correlation between the total stress and strain of the soil, so the stress-strain constitutive relationship of soils should and must be established on the variables of soil skeleton stress and soil skeleton strain.

As aforementioned, pore fluid pressure and effective stress have different effects on the strength and deformation of a soil. The pore fluid pressure is applied on the surfaces of soil particles. It only results in the volumetric deformation of soil particles and affects the shear strength on the interfaces of soil particles. When the volumetric change of soil particles caused by pore pressure is neglected, the stress-strain relationship of soils could be described by the relationship between soil skeleton strain and external forces-induced soil skeleton stress (effective stress). This relationship can be represented with the following formula:

$$\{\Delta\varepsilon\} = [D]^{-1}\{\Delta\sigma\} \tag{3.16}$$

where $\{\Delta\varepsilon\}$ is the matrix of strain increments, $\{\Delta\sigma\}$ is the matrix of stress increments and $[D]$ is the elasticity matrix of soils. When the volumetric change of the soil skeleton caused by pore fluid pressure needs to be considered, the relationship between the volumetric strain and soil skeleton stress can be expressed as:

$$\Delta\varepsilon_V = \delta_{ij}[C]\{\Delta\sigma\} + C_s\Delta u^* \tag{3.17}$$

where $[C]$ is the flexibility matrix of soils, C_s is the volumetric compression coefficient of soil particles.

In most cases, since $C_s \ll C$, and Δu^* is not very great in quantitative value, so the volumetric change caused by pore fluid pressure, i.e., $C_s\Delta u^*$, can be ignored.

It should be clarified that the elasticity matrix and flexibility matrix in Equations (3.15) and (3.17) are both relevant to stress state. However, the elasticity or flexibility matrix does not affect the state of internal force equilibrium of the soil, that is to say, it doesn't change the effective stress of a soil.

3.3 Primary verification of the correlation between effective stress and shear strength of unsaturated soils

The shear strength of a soil is that of the soil skeleton, which is determined by skeleton stress. The soil skeleton stress can be divided into effective stress and pore fluid pressure-induced soil skeleton stress, and the pore fluid pressure only contributes to the shear strength of the soil skeleton on particles' interfaces. Therefore, when this kind of contribution can be neglected, the effective stress governs the shear strength of a soil. Enormous experimental and empirical studies have indicated that the effect of pore water pressure on the strength and deformation of a saturated soil is negligible in most cases. Consequently, effective stress exclusively controls the strength and deformation of a soil. As it works for saturated soils, it should be applicable for unsaturated soils that effective stress controls its strength and deformation.

Based on the stress state variables for unsaturated soils developed by Fredlund and Morgenstern (1977), Vanapalli and Fredlund (Vanapalli *et al.*, 1996) proposed the shear strength equation of unsaturated soils with soil-water characteristic curve and shear strength parameters of saturated soils. The equation they have derived is the same as Equation (3.9).

Lu *et al.* (2010) proved the validity of using Equation (3.9) to predict the shear strength of unsaturated soils based on certain published experimental data available in the literature. Alonso *et al.* (2010) divided pore water into macro-pore occupied water and micro-pore occupied water. By treating the micro-pore water as immobile water irrespective of applied suction and mechanical load, they defined the effective stress equation by replacing the parameter χ in Bishop's equation with effective degree of saturation. They also proved that shear strength criterion based on that effective stress determines the shear strength and shear stiffness using some examples available from published literature.

A great number of researchers have reported the shear strength data of unsaturated soils, and they proposed various shear strength equations to predict the shear behavior for unsaturated soils (e.g., Alonso *et al.*, 2010; Hossain and Yin, 2010; Kayadelen *et al.*, 2007; Khalili and Khabbaz, 1998; Khalili *et al.*, 2004; Lee *et al.*, 2005; Miao *et al.*, 2015; Sheng *et al.*, 2011; Zhan and Ng, 2006). Here, some examples (Hossain and Yin, 2010; Khalili *et al.*, 2004; Miao *et al.*, 2015; Zhan and Ng, 2006) have been collected, and the shear strength data has been reanalyzed basing on the perspective that the effective stress controls the shear strength for unsaturated soils.

Khalili *et al.* (2004) conducted suction controlled triaxial tests on three types of undisturbed soil samples, which were all taken from the Hume Dam in Australia and designated as SJ10a, SJ10b and SJ11, respectively.

The saturated shear strength parameters c' and φ' were measured by traditional saturated triaxial tests. For saturated soils, the pore pressure inside the soil samples was measured during the shearing process. The tests of unsaturated soils were conducted by a triaxial testing instrument modified based on the Bishop-Wesley hydraulic triaxial cell. The samples had a diameter of 50 mm and a height of 100 mm. The net normal stress for all the samples was 200 kPa. The samples were tested by controlling the matric suction in a range of 100 kPa to 400 kPa through axial translation technique with a shearing rate of 0.003%/min. The air-entry values of SJ10a, SJ10b and SJ11 are 95 kPa, 125 kPa and 200 kPa, respectively.

The procedure of the triaxial test was as follows: the original prepared soil samples with the inside pore water pressure of 0 kPa were placed in the pressure chamber; after the chamber was filled with water, a confining pressure of 10 kPa was applied; then the saturated ceramic disk with high air-entry value at the bottom of the samples was connected to the atmosphere to drain the sample free of water. After that, the air pressure inside the samples was gradually increased to the target value of suction, and the confining pressure was increased simultaneously to keep it constantly 200 kPa higher than the air pressure for an identical value of net normal stress ($\sigma_3 - u_a = 200$ kPa). When the target value of air pressure and confining pressure was achieved, the desired suction was applied to the soil, and the specimen was left to equalize until the flow of water into or out of the specimen was extreme small. That is called the

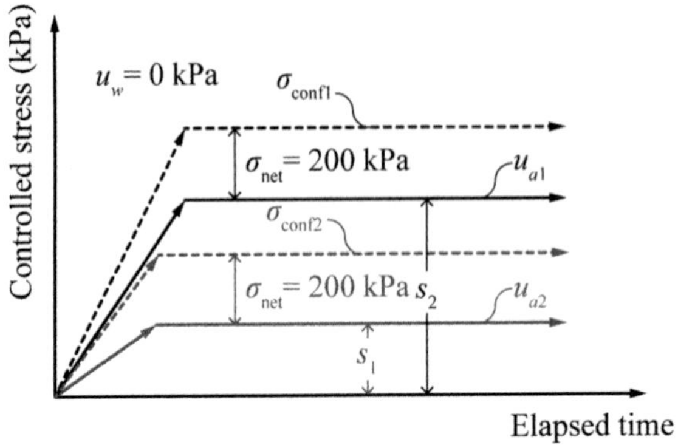

Figure 3.1 The procedure of applying suction and suction equalization

suction equalization stage. The stress change of the soil samples during the suction equilibration stage is shown in Figure 3.1.

The effective stress at the state when the suction reached equilibrium (before shearing) can be expressed as $\sigma_3' = \sigma_3 - u_a + S_e(u_a - u_w)$, where the effective stress σ_3' can be obtained by using the SWCC curve. After the suction equalization was achieved, the soil specimen was sheared to failure at a constant displacement rate. The suction is usually kept constant during the shearing process. For unsaturated soils, if the shear strength equation is established in the framework of effective stress (see Equation 3.9), the shear strength envelope line in effective stress versus shear strength stress space is consistent to that of saturated soils. Based on the Mohr-Coulomb strength principle, it can predict that the maximum deviatoric stress satisfies:

$$q_{\max} = \frac{2c'\cos\varphi'}{1-\sin\varphi'} + \frac{2\sin\varphi'}{1-\sin\varphi'}\sigma_3' \tag{3.18}$$

The relation of the effective stress σ_3' and maximum deviatoric stress $q_{\max}$ for the samples SJ10a, SJ10b and SJ11 obtained in the experiments is drawn in Figures 3.2, 3.3 and 3.4. The relation between σ_3' and $q_{\max}$ predicted with Equation (3.18) is also drawn in the figures with the solid line.

The results showed that the shear strength of unsaturated soil can be well predicted using the extended Mohr-Coulomb shear strength criterion on the basis of effective stress.

Zhan and Ng (2006) took unsaturated expansive silty clay from the South-to-North Water Transfer Canal in Hubei Province, China, for preparing undisturbed and compact unsaturated soil samples. They conducted suction-controlled direct shear tests on the samples to investigate the shear strength of the soil. Here, their experimental results have been reanalyzed to identify that effective stress controls the shear strength of unsaturated soil.

The dimension of the compact unsaturated soil samples was 50.5 mm × 50.8 mm × 21.4 mm. The compacted samples were prepared with the maximum compact stress of 800 kPa with water content of 18.5% (lower than the optimum water content 20.5% obtained in the Proctor compaction test). The obtained specimen had a dry density to the average value (i.e., 1.56 g/cm^3) of the natural specimens. The original samples had the same dimension to

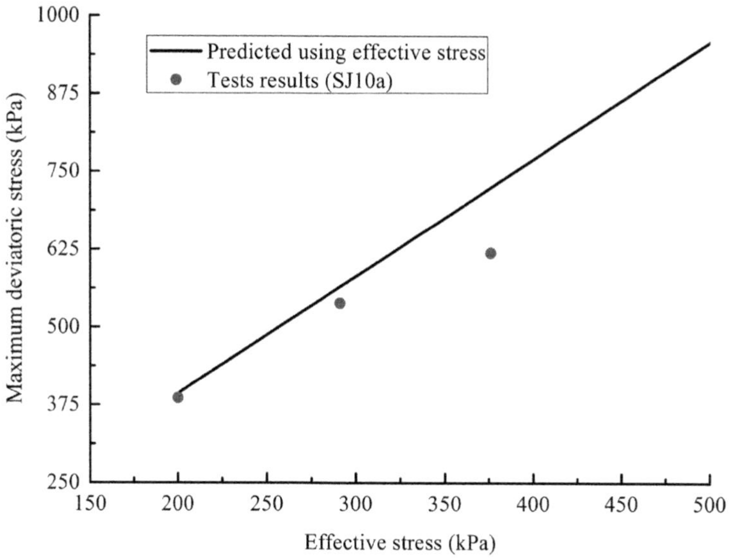

Figure 3.2 Effective stress and maximum deviatoric stress of sample SJ10a

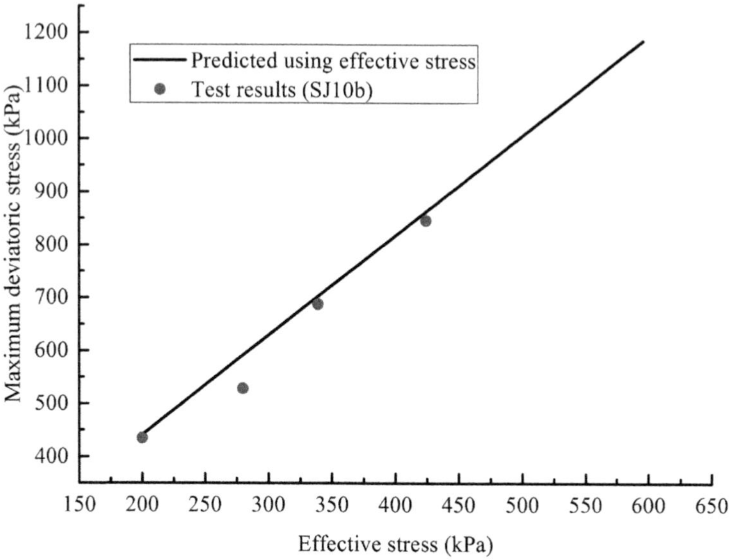

Figure 3.3 Effective stress and maximum deviatoric stress of sample SJ10b

the compact samples, with the initial water content in the range of 22.0% ~ 23.5% and initial dry density between 1.52 g/cm^3 and 1.56 g/cm^3.

With the peak strength obtained in the direct shear tests of original and compact saturated soil samples, the shear strength parameters of saturated soils can be obtained as: for original samples, c' = 16.2 kPa and φ' = 28.7°; and for compact samples, c' = 0 kPa and φ' = 24°.

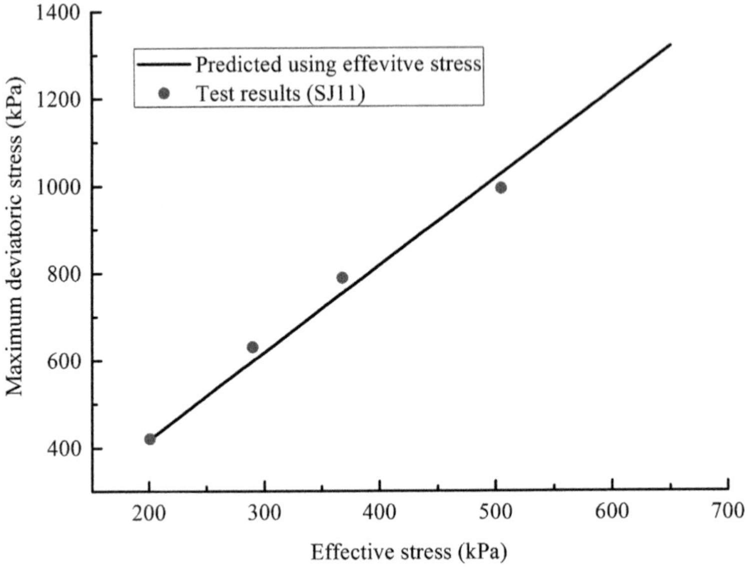

Figure 3.4 Effective stress and maximum deviatoric stress of sample SJ11

The direct shear testing of unsaturated soils was conducted by a modified direct shear testing instrument. A fully closed air pressure chamber was installed out of the traditional direct shear test instrument. Meanwhile, a saturated ceramic plate with the air-entry value of 500 kPa was installed at the bottom of the sample. The matric suction in the specimen was controlled by using the axis translation technique. In addition, the modified direct shear apparatus had two linearly displacement transducers for monitoring horizontal and vertical displacements, a load cell for measuring shear force, a pressure transducer for monitoring water pressure and the differential pressure transducer for measuring water volume.

The direct shear test of each soil sample included three steps: the suction equalization, compression and drained shearing at constant suction. The constant displacement rate was maintained at 0.0019 mm/min during the shearing until the horizontal displacement reached 6 mm.

Based on the shear strength equation using effective stress, contribution of the suction to the shear strength of unsaturated soils can be expressed as follows.

$$\tau_f = S_e(u_a - u_w)\tan\varphi' \tag{3.19}$$

As it is shown in Figure 3.5, the contribution of the suction to the shear strength for the compacted specimens shows an obvious nonlinear relationship. However, the original article did not analyze the relationship between shear strength and effective stress.

Based on the experimental data in the original literature, the relationship between effective stress and the shear strength can be obtained as it is shown in Figure 3.6. The solid line shows the prediction of the proposed shear strength equation, which indicates a good fit between the prediction and the test results of the shear strength. In addition, the relationship

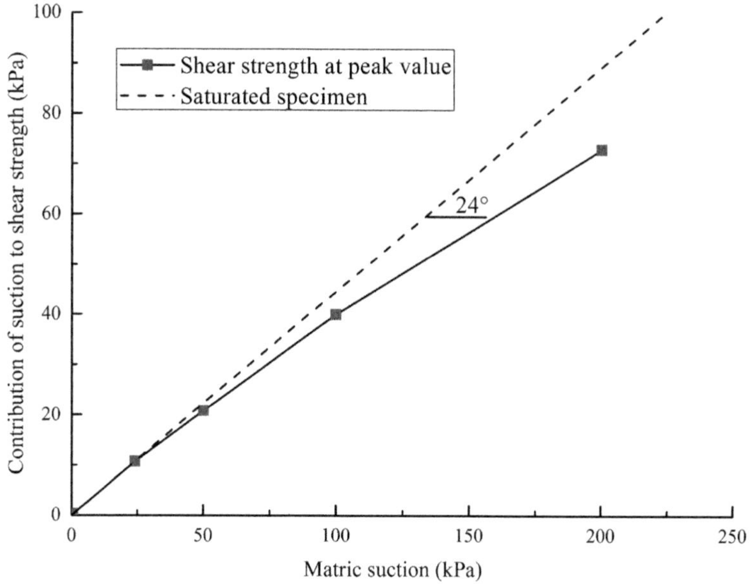

Figure 3.5 Contribution of suction to shear strength for compacted soil specimens

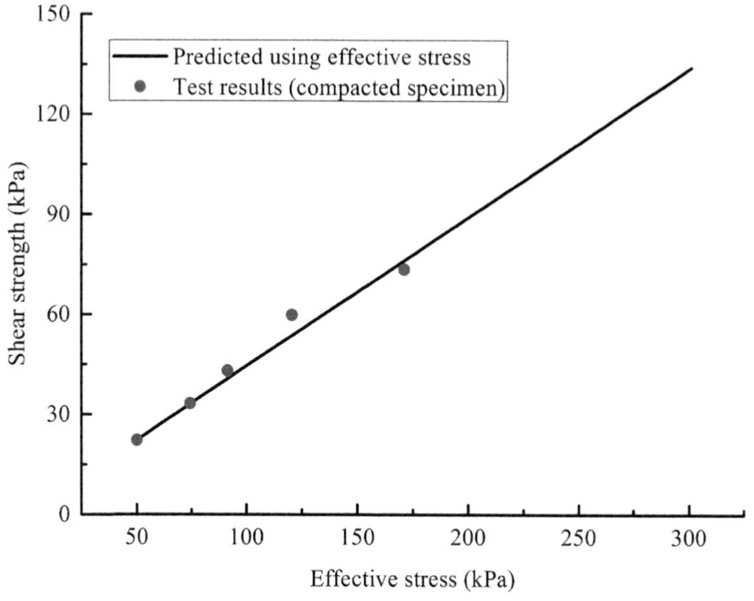

Figure 3.6 Comparison between predicted and measured unsaturated shear strength for compacted soil specimen

between suction and shear strength is shown in Figure 3.7, with the solid line presenting the prediction.

It is found from the curves of effective stress versus shear strength that (1) the extended Mohr-Coulomb shear strength principle can accurately predict the strength of unsaturated soils, which can be obtained in suction-controlled direct shear tests, and (2) the saturated soils and their counterpart of unsaturated soils satisfy the same strength principle, with the same shear strength parameters and the shear strength determined by effective stress, as is shown in Figure 3.6. There is an obvious air-entry value point in the SWCC of compacted soil samples. Therefore, there is an apparent turning in the matric suction and shear strength relationship curve, where the envelope starts to deviate from the failure envelope line of saturated soils, as illustrated in Figure 3.7.

Figure 3.8 presents the relationship between the effective stress and shear strength by taking the peak value as the shear strength of the soil, while Figure 3.9 shows the relationship between the matric suction and the shear strength. The solid line represents the predicted strength based on the shear strength criterion.

It can be seen that the shear strength criterion can provide a good prediction of the shear strength for the suction-controlled direct shear tests when the matric suction is 100 kPa, while the predicted results do not agree with the experimental results as well when the matric suction is 200 kPa. An overestimate can be seen when the suction of the specimen is 200 kPa. As is shown in the curves of shear strength and horizontal displacement (Zhan and Ng, 2006), the shear stress decreased obviously during the shear process when the specimen is under 200 kPa suction, and hence the real shear strength is much lower than the predicted value.

Miao *et al.* (2015) prepared samples at different degrees of saturation using a swelling soil from Nanning, Guangxi Province, China. Triaxial tests were conducted on these samples with different confining pressures, and the pore water inside the samples did not flow freely during the tests, which can be considered as triaxial tests of unsaturated soils with constant water contents.

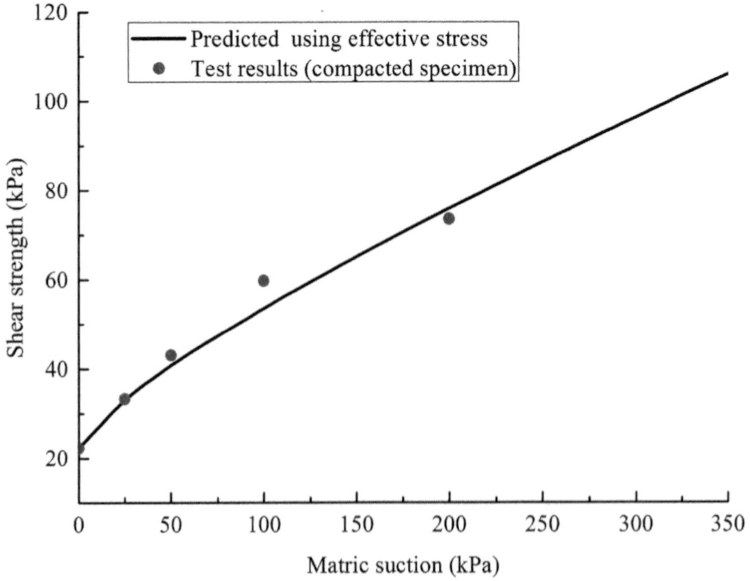

Figure 3.7 Suction and shear strength of compacted soil samples

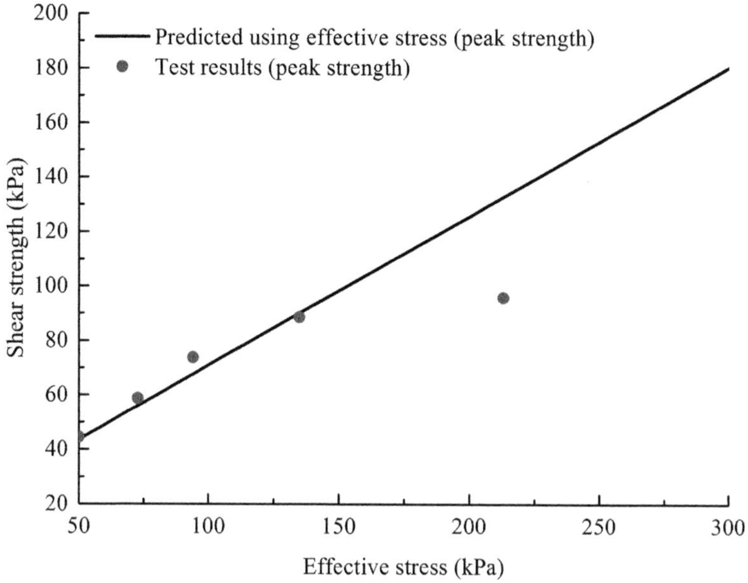

Figure 3.8 Effective stress and shear strength of the disturbed specimens

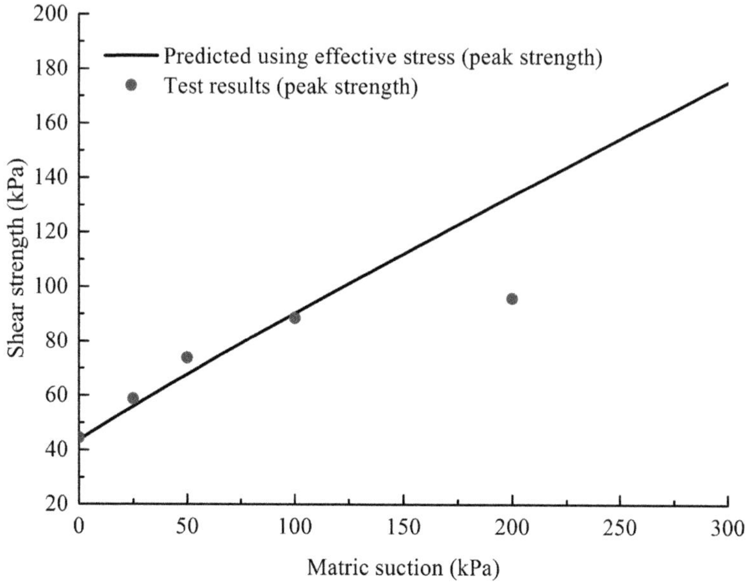

Figure 3.9 Matric suction and shear strength (the disturbed specimens)

Before the shearing of the soil samples, the degrees of saturation were initially controlled at 76.3%, 83.5%, 92.1% and 100%. The testing confining pressures were controlled at 50 kPa, 100 kPa and 200 kPa for each degree of initial saturation, respectively.

The experimental data shows that the degree of saturation of the samples changed slightly during the shearing process, indicating an ignorable volume change of the specimens. The SWCC of the swelling soil is drawn in Figure 3.10.

The curves of effective stress versus maximum deviatoric stress in addition to matric suction vs. maximum deviatoric stress are drawn in Figures 3.11 and 3.12, respectively.

Similar to the these verification results for different unsaturated soils, the shear strength for Guangxi swelling clay can be accurately predicted by the shear strength principle based on effective stress. The relationship between effective stress and shear strength satisfies the proposed shear strength principle, which is the same for saturated soils.

Hossain and Yin (2010) conducted suction-controlled direct shear tests on the CDG soil that is widely distributed in Hong Kong, which was investigated by many researchers. The particle-size distribution indicated that this type of soil was composed of 5.8% gravel, 44.1% sand, 36.8% silt and 13.3% clay, with the median diameter (D_{50}) of 0.063 mm. According to Unified Soil Classified System (ASTM1992), this type of soil is classified as silty sand. The plastic limit and liquid limit were 22.7% and 32.8%, respectively. The maximum dry density determined by the 2.5 kg hammer standard compaction tests was 1.75 g/cm^3, and the corresponding optimum water content was 14.33%. The specific gravity of the soil, G_s, was 2.599.

The SWCC of this soil is drawn in Figure 3.13. As it is shown in the figure, the air-entry value of the specimen is about 25 kPa. The curve fitting of the SWCC indicates that the residual degree of saturation of this soil is about 45%.

The procedure of the suction-controlled direct shear test was basically the same as that reported by Zhan and Ng. (2006). During the test, the matric suction was controlled at 0 kPa,

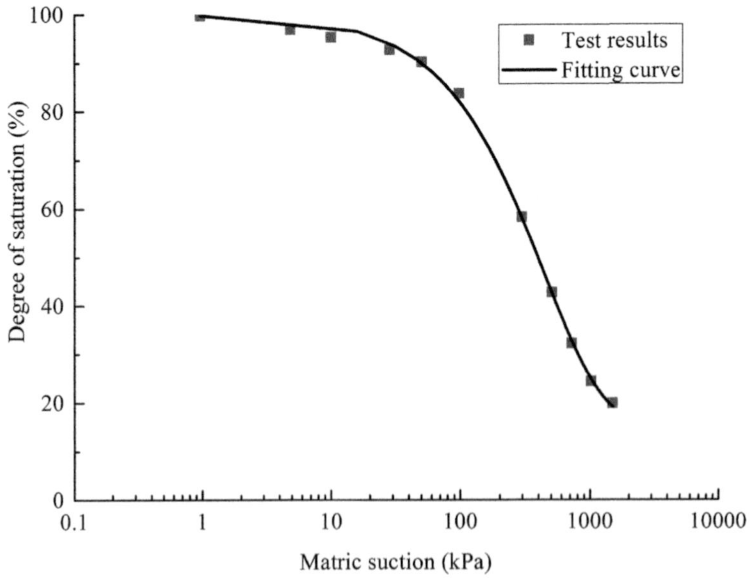

Figure 3.10 SWCC of the swelling soil from Guangxi Province

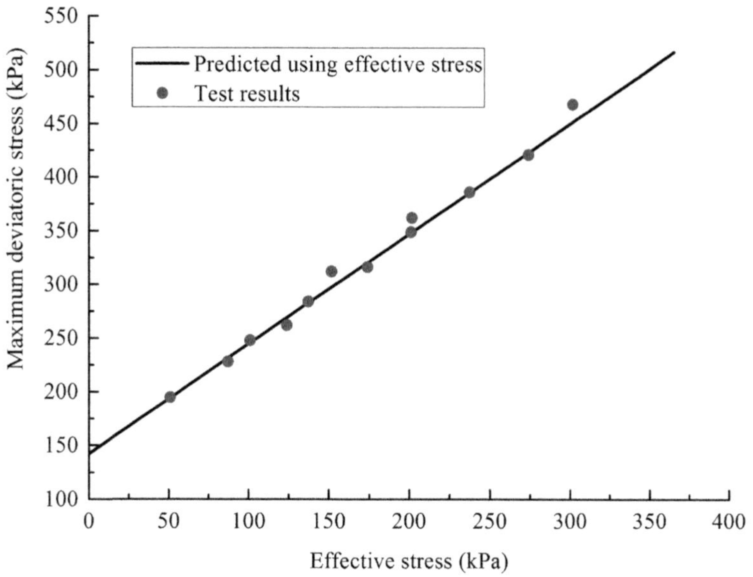

Figure 3.11 Effective stress versus maximum deviatoric stress for Guangxi swelling clay

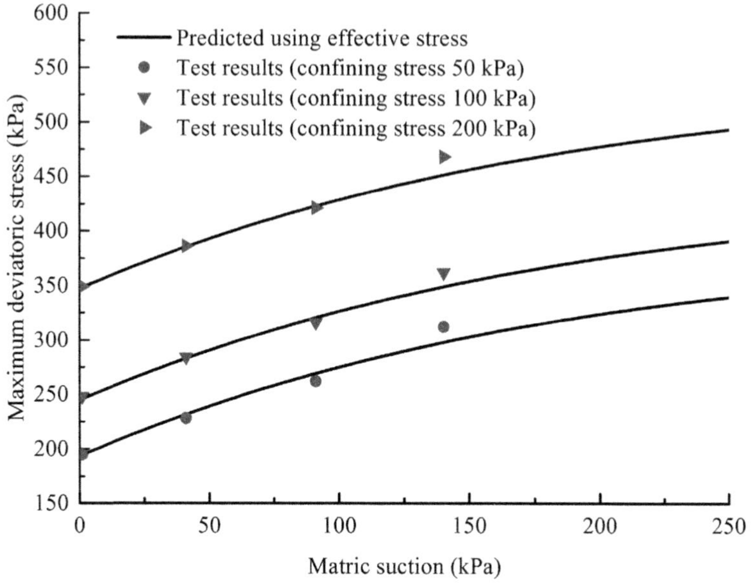

Figure 3.12 Matric suction versus maximum deviatoric stress for Guangxi swelling clay

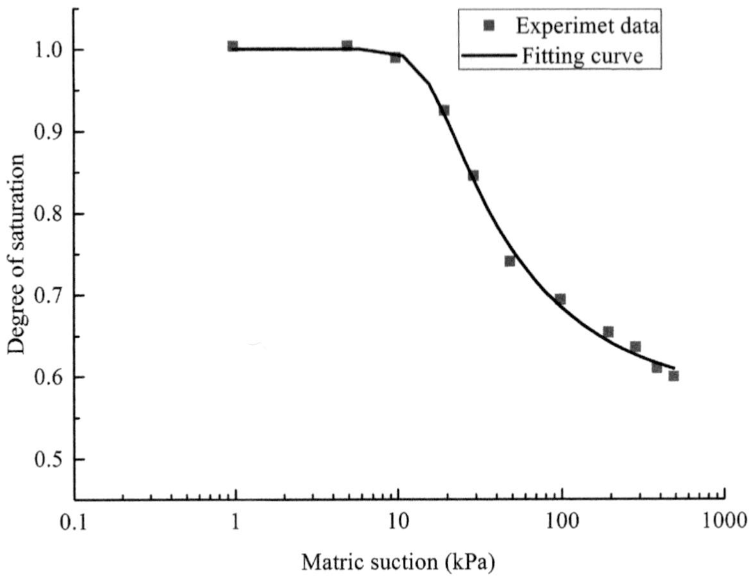

Figure 3.13 SWCC of CDG soils from Hong Kong

50 kPa, 100 kPa, 200 kPa and 300 kPa, with the corresponding net normal stresses of 50 kPa, 100 kPa, 200 kPa and 300 kPa, respectively. In total, 20 groups of tests were conducted in the testing program. The vertical deformation and water movement of the samples were measured during the tests simultaneously. The testing results illustrated that the soil samples under high suction would present certain dilatancy behavior in the same confining pressure conditions. This dilatancy behavior gradually decreased with the increase of net normal stress. The shear strength parameters of this soil obtained through the direct shear test at saturated state (i.e., matric suction = 0 kPa) were $c' = 0$ kPa, $\varphi' = 29.9°$. The tests with the net stress of 50 kPa and 100 kPa are analyzed herein based on the SWCC. The shear strength predicted with the shear strength equation of unsaturated soils (Equation (3.9)) and the measured shear strength are compared in Figure 3.14.

For the two groups of test results (net stress of 50 kPa and 100 kPa) analyzed, the shear strength equation based on effective stress can accurately predict the shear strength of the CDG soils, except for a few cases that underestimate or overestimate the strength of the soil. In addition, the relationship between effective stress and shear strength satisfies the Mohr-Coulomb shear strength equation for saturated soils. To investigate the influence of matric suction on shear strength, the relationship between suction and shear strength under different net normal stress is also presented in Figure 3.15.

Since the prediction of shear strength in different net stress conditions was conducted based on the same SWCC, which was tested without any load, the suction-shear strength envelope has the same shape, which is translated along the axial of shear strength, as indicated in Figure 3.15. Similar to the experimental results by the other researchers, it presents an apparent nonlinear relationship between the matric suction and shear strength under specific net normal stress. In the suction range of 0 to 300 kPa, the strength envelope approximately

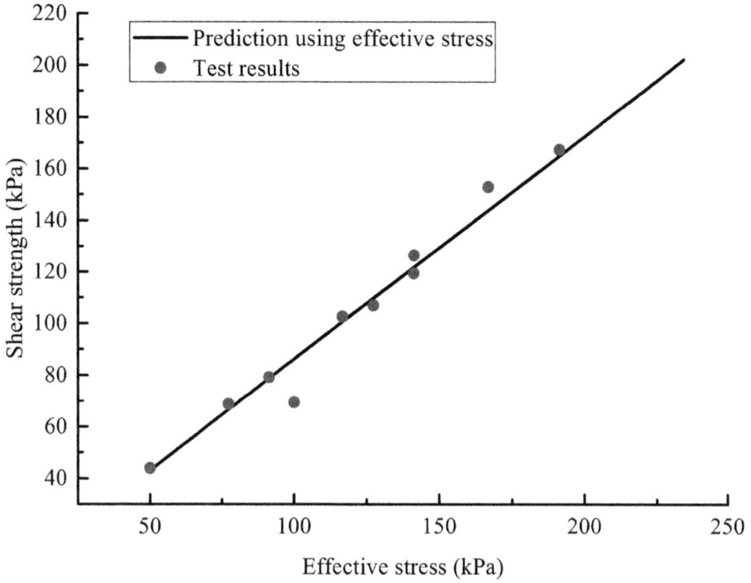

Figure 3.14 The comparison between the predicted and measured shear strength for the CDG soils from Hong Kong

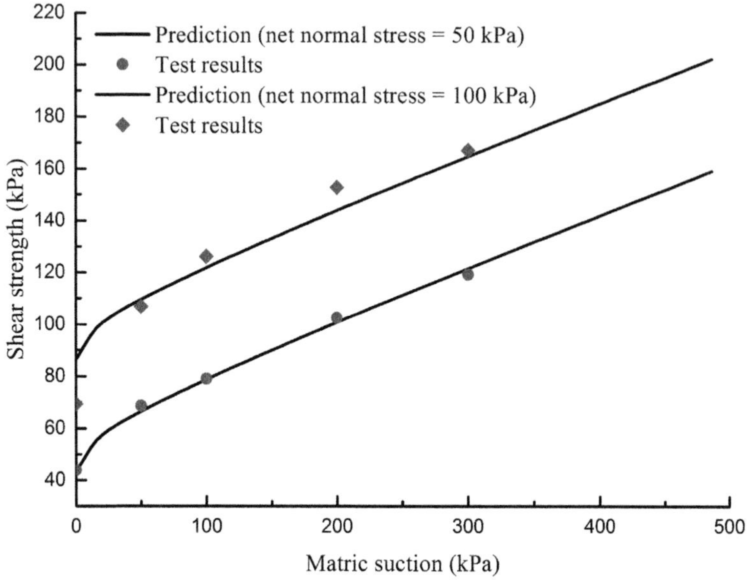

Figure 3.15 The comparison between the predicted and measured shear strength for the CDG soils from Hong Kong

presents a bilinear property. This bilinear line appears to have an apparent turning at the air-entry value of 25 kPa. This is consistent with the study results by many other researchers (Khalili and Khabbaz, 1998; Khalili *et al.*, 2004).

Another group of data is from Kayadelen *et al.* (2007). In that research, a type of residual soil formed by the alteration of basalts and basaltic tuff in the Diyarbakir area in Turkey was investigated. Since the soil samples were collected from a semi-arid area, soils experienced an accelerated evaporation process during the dehydration process. Therefore, there were many fine cracks in the natural soil, some of which were even visible to the naked eye. The plastic limit and liquid limit were 77% and 32%, respectively, and the corresponding plastic index was 46%. The SWCC of this soil is shown in Figure 3.16. The air-entry value of this soil is about 40 kPa.

The shear strength parameters of this soil at saturated state obtained with conventional triaxial tests were $c' = 14.82$ kPa, $\varphi' = 21.9°$. Kayadelen *et al.* conducted 12 groups of triaxial tests in which the matric suction was controlled at different values ranging from 50 to 400 kPa, while the corresponding net normal stress was 50 kPa, 150 kPa, and 200 kPa, respectively.

Kayadelen *et al.* distinguished the effects of net stress and matric suction on the shear strength and investigated the relationship between shear strength and net normal stress. The results illustrated that matric suction barely had any effect on the internal fictional angle φ' because the unsaturated shear strength envelopes are approximately parallel to the saturated shear strength envelope. Meanwhile, the contribution of matric suction to the shear strength has been studied, and this contribution varies non-linearity with respect to matric suction that was similar to the other researchers' results. The results showed that the suction-shear

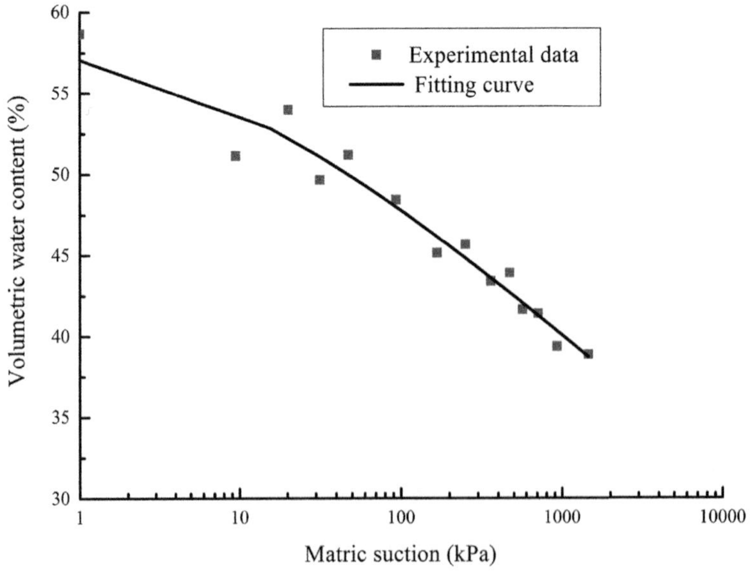

Figure 3.16 Soil-water characteristic curve of Diyabarkir residual soils (after Kayadelen *et al.*, 2007)

strength envelope became smoother when the air-entry value was exceeded. As presented in Figure 3.17, the relationship between effective stress and maximum deviatoric stress has been established as the same approach described previously. According to the results, it was found that effective stress and shear strength satisfy the Mohr-Coulomb shear strength principle of saturated soils.

Data analysis on the results from all these shear strength tests of unsaturated soils illustrates that the shear strength of unsaturated soils can be predicted using the shear strength parameters of the soil in saturated state and the effective stress, and the proposed shear strength equation was in satisfactory agreement with the experimental results.

SWCC was used in the shear strength prediction in the above examples. However, two points should be noted: (1) the change in the state of soils would lead to the change of SWCC during the deformation of soils; and (2) the shear strength of unsaturated soils predicted with SWCC that is measured without confining stress is not applicable for all conditions, especially for accurate prediction. This is because that SWCC represents the relationship between matric potential (matric suction) and water content (degree of saturation) of soils under hydrostatic equilibrium state. When pore water moves in or out of the soil, the relationship between water content (degree of saturation) and matric potential would not follow the SWCC. So for accurate prediction, the seepage equation of unsaturated soils needs to be solved to determine the water content (degree of saturation) and corresponding matric suction when the soil is not under hydrostatic equilibrium state. Furthermore, stress state influences the density of a soil, and then the SWCC. As the SWCC is generally obtained by testing a soil sample in self-gravity without additional stress, first the effect of stress state on density and then the SWCC should be taken into account for accurate prediction of the shear strength.

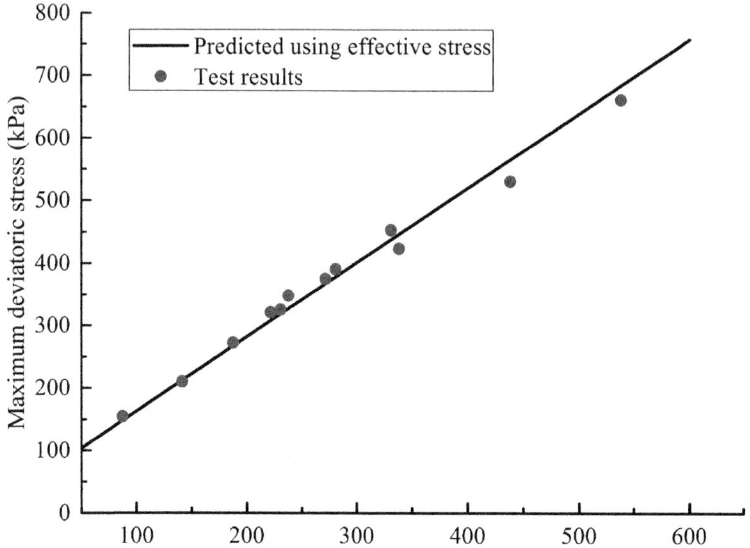

Figure 3.17 Correlation curve of effective stress vs. maximum shear stress of Diyabarkir residual soils from Turkey (the data was collected from the test results by Kayadelen *et al.*, 2007)

On the other hand, the equation of effective stress of unsaturated soils is always correct no matter what the water seepage (change of water content) and the change of matric suction are. Therefore, the key to determine shear strength of unsaturated soils is to identify the water content and matric suction at the current state of the soil. In fact, it is necessary to identify the effective stress for the shear strength prediction, and the key is to determine the water content and matric suction (more accurately, water pressure and air pressure). Consequently, the determination of water content and soil-water potential is the key to determine the shear strength of unsaturated soils.

However, when the absolute water content is lower than the soil skeleton water content (which is assumed as the residual water content for the moment), the soil becomes a kind of dry soil. The effective stress is the difference between total stress and pore air pressure. Although we can still believe that the effective stress controls the strength and deformation of the soil, in this case, the properties of shear strength and deformation, i.e., the strength and deformation parameters of the soils, may be totally different from those at saturated state and may change with the absolute water content.

Additionally, it should be noted that the SWCC has lost its effectiveness when the absolute water content of the soil is lower than the residual water content.

3.4 Effective stress principle for unsaturated soils

3.4.1 Statement of effective stress principle

Terzaghi's effective stress principle had been stated by Jennings and Burland (1962) in the form of two propositions: (1) changes in volume and shearing strength of a soil are due exclusively to changes in effective stress; (2) the effective stress in a soil is defined as the excess of the total applied stress over the pore pressure.

Terzaghi's effective stress principle is very important for saturated soil mechanics, of which the application is very successful. Similarly, it is always desirable to search for an effective stress principle that is feasible for unsaturated soils. It can be well known from the derivation of effective stress equation that the strength and deformation of unsaturated soils are controlled by effective stress, as the effect of pore pressure can be ignored. Since the strength and deformation of a soil are those of skeleton and determined by skeleton stress, a uniform effective stress principle that is applicable for a soil in both saturated and unsaturated states may be stated as:

(1) Effective stress is the soil skeleton stress induced by all the external forces, excluding pore fluid pressure;
(2) The relationship between the effective stress, the total stress and pore fluid pressure satisfies $\sigma' = \sigma_t - u_a + S_e(u_a - u_w)$. This is the effective stress equation for a soil in both saturated and unsaturated states;
(3) The shear strength and deformation are determined by effective stress when the effect of pore fluid pressure on the shear strength and deformation can be neglected.

Among these statements, term (1) is introducing the physical meaning of effective stress; term (2) formulates the form of effective stress; and term (3) is the applicable conditions that uses the effective stress to determine the strength and deformation of a soil.

In fact, when the soil skeleton is clarified to be the research subject and the effect of pore fluid pressure is taken into account separately, the effective stress principle does not seem to be of much significance. In this case the effective stress equation is a natural result of force and equilibrium analysis of the free body of a soil.

On the other hand, as will be mentioned in Chapter 5, the effective stress equation of unsaturated soils is actually a specific combination of stress state variables, which can be the combination of net stress and matric suction or the combination of total stress, pore water pressure and matric suction, as long as the combination form and the corresponding parameters are determined. Therefore, the effective stress principle approach and the stress state variable approach for unsaturated soils are not in conflict to each other. In a broader sense, the stress state variables can describe the effect of each stress state variable separately. However, the stress state variables approach cannot identify exactly which kind of stress combination controls the strength and deformation.

In geotechnical engineering, the total stress of soils is easier to determine in most cases. If it is also easy to identify the pore water pressure and pore air pressure, the effective stress can easily be calculated by using the effective stress equation. Then the mechanical behavior of unsaturated soils can be determined by effective stress.

3.4.2 Simple applications of effective stress principle for unsaturated soils

The core of effective stress principle is actually the effective stress equation, which is represented by the relationship between the total stress, effective stress and pore fluid pressure. For any soil layers, the total stress can be directly calculated, and the effective stress can be obtained using the effective stress equation directly as long as the pore water pressure and pore air pressure are known. Therefore, determining the pore fluid pressure (i.e., water pressure and air pressure, similarly hereinafter) distribution is the key to determining the effective stress of soil layers. In unsaturated soils, the water content usually stays at a varying state, and hence the calculation of total stress and pore fluid pressure distribution reflects the transient response of the soil, so the corresponding effective stress distribution is a transient behavior.

In the following, the application of the effective stress equation with an example of a homogeneous unsaturated soil layer will be demonstrated. For simplicity, the soil is assumed homogeneous in all the following examples, of which the stress-strain relationship is linear elastic, which follows Hook's law.

Example 1. The stress in unsaturated soil layers without seepage

The unsaturated soil layer above the underground water level is shown in Figure 3.18. It is assumed that the soil layer is open to the atmosphere and without any seepage. The effect of temperature change is not considered. The water level is assumed to be kept at a fixed height.

According to the equilibrium differential equation in total stress, we have:

$$\frac{d(\sigma_{tz})}{dz} = -\rho_{(z)}\,g \tag{3.19}$$

where ρ is the natural density of unsaturated soils. The boundary condition is $\sigma_{tz} = 0$ at $z = H$.

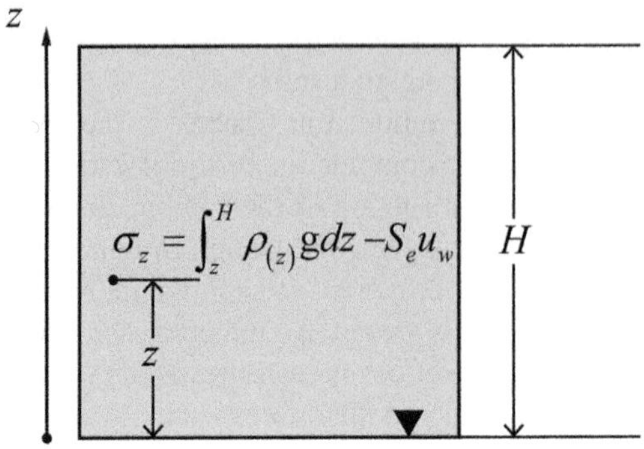

Figure 3.18 Effective stress distribution in unsaturated soil layer above water

Accordingly, the solution of Equation (3.19) is:

$$\sigma_{tz} = \int_z^H \rho_{(z)}\, g\, dz \tag{3.20}$$

Next the effective stress equation of unsaturated soils is applied. Since the soil layer is open to the atmosphere, $u_a = 0$, then:

$$\sigma_z = \sigma_{tz} - S_e u_w = \int_z^H \rho_{(z)}\, g\, dz - S_e u_w \tag{3.21}$$

The same result can be obtained with the equilibrium differential equation in terms of effective stress. It can be represented as follows:

$$\frac{d(\sigma_z + S_e u_w)}{dz} = -\rho_{(z)}\, g \tag{3.22}$$

The boundary conditions are: $\sigma_z + S_e u_w = 0$ at $z = H$, and $u_w = 0$ at $z = 0$.

The general solution for these differential equation can be written as:

$$\sigma_z + S_e u_w = -\int_a^z \rho_{(z)}\, g\, dz + C, \quad a \in (0, H) \tag{3.23}$$

Substituting the boundary conditions listed earlier, and thus:

$$\sigma_z = \int_z^H \rho_{(z)}\, g\, dz - S_e u_w \tag{3.24}$$

Applying the effective stress equation in a horizontal direction, the equation is:

$$\sigma_x = \sigma_y = \sigma_{tx} - S_e u_w \tag{3.25}$$

The relationship between the effective stress and strain of the soil satisfies the generalized Hook's law:

$$\varepsilon_x = \frac{1}{E}(\sigma_x - \nu\sigma_x - \nu\sigma_z) \tag{3.26}$$

$$\varepsilon_x = \frac{1}{E}\left\{\sigma_{tx} - S_e u_w - \nu\left[\sigma_{tx} - S_e u_w\right] - \nu\left[\int_Z^H \rho_{(z)}\, g\, dz - S_e u_w\right]\right\} \tag{3.27}$$

$$\varepsilon_x = \frac{1}{E}\left[(1-\nu)\sigma_{tx} - (1-2\nu)(S_e u_w) - \nu\int_Z^H \rho_{(z)}\, g\, dz\right] \tag{3.28}$$

The following solution can be obtained while $\varepsilon_x = 0$:

$$\sigma_{tx} = \frac{1-2\nu}{1-\nu}(S_e u_w) + \frac{\nu}{1-\nu}\int_z^H \rho_{(z)}\, g\, dz \tag{3.29}$$

Therefore:

$$\sigma_x = \sigma_y = \sigma_{tx} - S_e u_w = K_0\left(\int_z^H \rho_{(z)}\, g\, dz - S_e u_w\right) \tag{3.30}$$

where $K_0 = \dfrac{\nu}{1-\nu}$ is the horizontal earth pressure coefficient for saturated soils.

It can be seen that the key to solving the effective stress of soil layers is to identify the pore water pressure. When there is no water flowing (seepage), the pore water pressure distribution in the soil layer is determined with water content (degree of saturation), which can be obtained using SWCC.

Example 2. Stress in homogeneous soil layer during the declining of underground water level

Let's consider a broad homogeneous soil layer that is open to the atmosphere. The water level declines from the ground surface to the underground location at the depth of H, then remains constant. The soil layer above the underground water level turned from a full-saturated state to an unsaturated state. The water content (degree of saturation) distribution in the soil layer is assumed to be known, and the seepage force was not considered. What follows shows the stress distribution of the soil layer under different conditions.

When the soil layer is fully saturated, the stress is:

$$\sigma_{z0} = \rho' g z \tag{3.31}$$

$$\sigma_{x0} = \sigma_{y0} = k_0 \sigma_{z0} \tag{3.32}$$

When it turns to a steady unsaturated state from a saturated state, the stress of the soil layer is:

$$\sigma_z = \int_z^H \rho_{(z)}\, g\, dz - S_e u_w \tag{3.33}$$

$$\sigma_x = \sigma_{tx} - S_e u_w \tag{3.34}$$

The density of soils turns to unsaturated natural density from buoyant density, i.e.:

$$\Delta\rho = (\rho_m - \rho_w n_a) - \rho' = \rho_w(1 - n + n_w) \tag{3.35}$$

Therefore:

$$\Delta\sigma_z = \sigma_z - \sigma_{z0} = \int_z^H \rho_w g(1-n+n_w)dz - S_e u_w \tag{3.36}$$

$$\Delta\sigma_x = \sigma_x - \sigma_{x0} = \sigma_{tx} - S_e u_w - K_0\sigma_{z0} = \Delta\sigma_{tx} - S_e u_w \tag{3.37}$$

$$\Delta\sigma_{tx} = \sigma_{tx} - K_0\sigma_{z0} \tag{3.38}$$

Applying the increments to the stress-strain constitutive relationship, it has:

$$\Delta\varepsilon_z = \frac{1}{E}\left\{\left[\int_z^H \rho_w g(1-n+n_w)dz - S_e u_w\right] - 2v(\Delta\sigma_{tx} - S_e u_w)\right\} \tag{3.39}$$

$$\begin{aligned}\Delta\varepsilon_x &= \Delta\varepsilon_y \\ &= \frac{1}{E}\left\{(\Delta\sigma_{tx} - S_e u_w) - v(\Delta\sigma_{tx} - S_e u_w) - v\left[\int_z^H \rho_w g(1-n+n_w)dz - S_e u_w\right]\right\}\end{aligned} \tag{3.40}$$

$$\begin{aligned}\Delta\varepsilon_V &= \Delta\varepsilon_x + \Delta\varepsilon_y + \Delta\varepsilon_z \\ &= \frac{1}{E}\left\{(1-2v)\left[\int_z^H \rho_w g(1-n+n_w)dz\right] - 3(1-2v)S_e u_w + 2(1-2v)\Delta\sigma_{tx}\right\}\end{aligned} \tag{3.41}$$

$$\Delta\varepsilon_V = \frac{1-2v}{E}\left\{\left[\int_z^H \rho_w g(1-n+n_w)dz\right] - 3S_e u_w + 2\Delta\sigma_{tx}\right\} \tag{3.42}$$

If $\Delta\varepsilon_x = \Delta\varepsilon_y = 0$, then

$$(1-v)\Delta\sigma_{tx} - (1-2v)S_e u_w - v\cdot\left[\int_z^H \rho_w g(1-n+n_w)dz\right] = 0 \tag{3.43}$$

$$\Delta\sigma_{tx} = \frac{v\cdot\left[\int_z^H \rho_w g(1-n+n_w)dz\right] + (1-2v)S_e u_w}{1-v} \tag{3.44}$$

Equation (3.44) represents the lateral total stress of the soil layer when the underground water level declines to a certain height and remains unchanged; the soil layer above the free water level turns from a saturated to an unsaturated state, without any deformation in horizontal direction.

Since:

$$\Delta\sigma_{tx} = \sigma_{tx} - K_0\sigma_{z0} \tag{3.45}$$

$$\sigma_{tx} = \Delta\sigma_{tx} + K_0\sigma_{z0} \tag{3.46}$$

Let $\sigma_{tx} = 0$, and substitute equation (3.44) into equation (3.46), then:

$$\frac{v\cdot\left[\int_z^H \rho_w g(1-n+n_w)dz\right] + (1-2v)S_e u_w}{1-v} + K_0\sigma_{z0} = 0 \tag{3.47}$$

$$u_w = -\frac{1}{(1-2v)S_e}\left\{(1-v)K_0\sigma_{z0} + v\cdot\left[\int_z^H \rho_w g(1-n+n_w)dz\right]\right\} \tag{3.48}$$

The lateral total stress of the unsaturated soil layer gradually decreases to 0. This means that soil lost the applied force in lateral directions at this point; it also means that the soil has reached the critical state to show tensile cracks. Therefore, Equation (3.48) is the condition for soil to reach the critical state of tensile cracks.

Example 3. The stress and strain of the sample in a confined compression test with free draining condition at top and bottom

Figure 3.19 presents a thin unsaturated soil sample in the confined compression test, with free drainage at top and bottom, i.e., $u_a = 0$. The effect of water flow (seepage) is neglected. When a loading is applied on the top surface of the sample:

$$\Delta\sigma_{tz} = \Delta P \tag{3.49}$$

$$\Delta\sigma_z = \Delta\sigma_{tz} - \Delta(S_e u_w) \tag{3.50}$$

i.e.,

$$\Delta\sigma_z = \Delta\sigma_{tz} - u_w \Delta S_e - S_e \Delta u_w \tag{3.51}$$

According to SWCC, $u_w = u_{w(S_e)}$ we can obtain $\Delta u_w = u_w{'}_{(S_e)} \cdot \Delta S_e$. Substituting this into equation (3.51):

$$\Delta\sigma_z = \Delta\sigma_{tz} - \Delta S_e[u_w + S_e u_w{'}] \tag{3.52}$$

i.e.:

$$\Delta\sigma_z = \Delta P - \Delta S_e[u_w + S_e u_w{'}] \tag{3.53}$$

in lateral direction (radial direction):

$$\Delta\sigma_r = \Delta\sigma_{tr} - \Delta(S_e u_w) \tag{3.54}$$

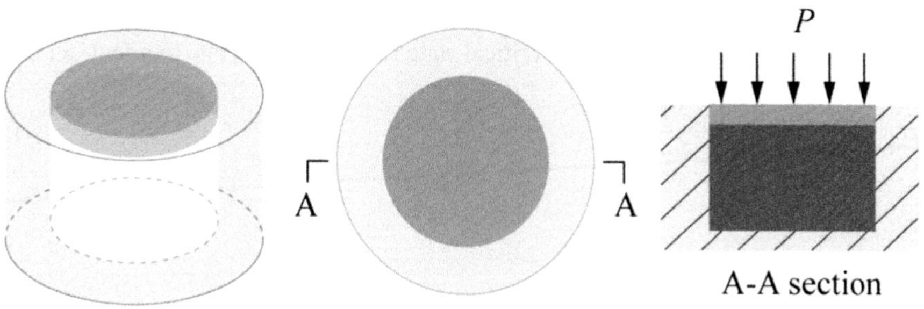

Figure 3.19 The sketch of a uniform soil sample in a stiff cylindrical chamber

and:

$$\Delta\varepsilon_r = \frac{1}{E}\left(\Delta\sigma_r - \nu\Delta\sigma_r - \nu\Delta\sigma_z\right) \tag{3.55}$$

$$\Delta\varepsilon_r = \frac{1}{E}\left\{\Delta\sigma_{tr} - \Delta\left(S_e u_w\right) - \nu\left[\Delta\sigma_{tr} - \Delta\left(S_e u_w\right)\right] - \nu\left[\Delta P - \Delta\left(S_e u_w\right)\right]\right\} \tag{3.56}$$

$$\Delta\varepsilon_r = \frac{1}{E}\left[(1-\nu)\Delta\sigma_{tr} - (1-2\nu)\Delta\left(S_e u_w\right) - \nu\Delta P\right] \tag{3.57}$$

Since $\Delta\varepsilon_r = 0$, then:

$$\Delta\sigma_{tr} = \frac{1-2\nu}{1-\nu}\Delta\left(S_e u_w\right) + \frac{\nu}{1-\nu}\Delta P \tag{3.58}$$

$$\Delta\sigma_{tr} = \frac{\nu}{1-\nu}\Delta P + \frac{1-2\nu}{1-\nu}\Delta S_e[u_w + S_e u_w{}'] \tag{3.59}$$

When the soil sample is shrinking and starts to separate from the chamber wall, $\sigma_{tr} = 0$, a critical condition can be obtained from the derivation as follows:

$$\sigma_{tz} = P \tag{3.60}$$

$$\sigma_z = \sigma_{tz} - (S_e u_w) = P - (S_e u_w) \tag{3.61}$$

In radial direction:

$$\sigma_r = \sigma_{tr} - \left(S_e u_w\right) \tag{3.62}$$

$$\varepsilon_r = \frac{1}{E}\left(\sigma_r - \nu\sigma_r - \nu\sigma_z\right) \tag{3.63}$$

$$\varepsilon_r = \frac{1}{E}\left\{\sigma_{tr} - \left(S_e u_w\right) - \nu\left[\sigma_{tr} - \left(S_e u_w\right)\right] - \nu\left[P - \left(S_e u_w\right)\right]\right\} \tag{3.64}$$

$$\varepsilon_r = \frac{1}{E}\left[(1-\nu)\sigma_{tr} - (1-2\nu)\left(S_e u_w\right) - \nu P\right] \tag{3.65}$$

When the soil sample starts to separate from the chamber wall, the interface between the soil sample and the chamber wall stays the critical state of cracking. At this moment, with $\varepsilon_r = 0$:

$$\sigma_{tr} = \frac{1-2\nu}{1-\nu}\left(S_e u_w\right) + \frac{\nu}{1-\nu}P \tag{3.66}$$

Let $\sigma_{tr} = 0$, we have:

$$\left(S_e u_w\right) = -\frac{\nu}{1-2\nu}P \tag{3.67}$$

Equation (3.67) represents the condition at which a soil sample starts to separate from a stiff cylinder chamber wall because of shrinkage.

Chapter 4

Seepage equation of unsaturated soils

The flow of fluid through porous medium is called seepage. Seepage in soil is referred to as the flow of pore water or pore air through the pores. The differential equation used to solve seepage problems is called the seepage equation. We now demonstrate the deduction of the seepage equation by taking the seepage of pore water for an example.

Traditionally, Darcy's law (1856) and the mass conservation equation of pore fluid are necessary in the derivation of the seepage equation (Braja, 2008; Lei *et al.*, 1988; Yin *et al.*, 2007). The formula of Darcy's law illustrates the relationship between the hydraulic potential gradient and seepage velocity, which is usually called the motion equation of pore water. For saturated soils, this relation is obtained via experiments. For unsaturated soils, this relation is derived by directly expanding the saturated soils' counterpart, which also can be obtained in experiments. In this chapter, we directly derive the formula of Darcy's law based on the equilibrium equation with the assumption that seepage resistance is proportional to seepage velocity. Subsequently, we incorporate this formula with the mass conservation condition of pore water and derive the seepage velocity equation. For the convenience of understanding, we primarily derive the seepage velocity equation of saturated soils.

4.1 Seepage equation of saturated soil

4.1.1 Derivation of the formula of Darcy's law for saturated soil

Since its velocity is small in general, the kinetic energy of pore water is negligible. We only consider its potential energy, i.e., the soil-water potential that drives pore water to flow. For saturated soils, soil-water potential includes gravity potential and pressure potential. In hydraulics, soil-water potential is also called water head. Gravity potential is called gravity head, or position head, and pressure potential is called pressure head. The sum of these two potentials is called total water head and also called piezometer head. The pore water flows from high water head towards low water head, overcoming soil resistance (soil-water interaction).

The formula of Darcy's law is the motion equation of pore water in seepage. For simplicity, we herein directly derive this motion equation with the equilibrium equation of pore water in the case of statics problems. The force of soil-water interaction is discussed first since it is included in the equilibrium equation of pore water as a component.

4.1.1.1 Soil-water interaction in saturated soils

There is an interaction force between the soil skeleton and the pore water. The pore water encounters resistance during its permeation through the soil skeleton and at the same time applies forces on the soil skeleton. The resistance that the water bears during its permeation is the force of soil-water interaction and also the seepage force, introduced later in this chapter.

In hydraulics, the flowing status of water is classified as laminar flow and turbulent flow. Most of the pore water flow in soil is laminar flow. The moving resistance on the laminar flow water is proportional to velocity.

Since the sectional area and shape of the pores in soil are irregular, the flowing of water in soil pores is complicated. Even for the pore water in sandy soils, the regulation of flow velocity distribution and true values of velocity are unable to be investigated. Therefore, we still need to apply the concept of physical quantities equalization at a point (Representative Elementary Volume, *REV*). In this case, we use the flux passing through unit area at unit time instant within the *REV* of soils to describe seepage velocity, also called permeation velocity and denoted by v.

The relationship between the full section average flow velocity at a point in soil v and the flow velocity at porous area v' can be determined by the following formula:

$$v = nv' \tag{4.1}$$

where n is the porosity of the soil.

For saturated soils, it can be assumed that the force of soil-water interaction is proportional to the velocity of pore water based on the research on the laminar flow resistance in hydraulics and the results of Darcy's permeation experiments, when pore water is in laminar flow status. Therefore, in unit soil mass, the interaction force applied on pore water can be expressed as:

$$f_i = \frac{a}{n} v_i \tag{4.2}$$

where f_i is the force of soil-water interaction, v_i is the velocity of pore water at the full section and a is the coefficient of the interaction force between pore water and the soil skeleton in the unit volume of the soil body.

For the deduction of the seepage equation for unsaturated soils, we use a_w to represent the factor of soil-water interaction force for the unit volume of pore water, not the unit volume of soil body. For saturated soils, $a_w = a/n$, i.e., $a = na_w$; and for unsaturated soils, $a_w = a/n_w$, i.e., $a = n_w a_w$.

According to the correlation between a and a_w, Equation (4.2) can be written as:

$$f_i = a_w v_i \tag{4.3}$$

4.1.1.2 Deduction of the formula of Darcy's law

On the condition that the soil is homogeneous and the inertia force of pore water can be ignored, we have the equilibrium differential equations of pore water as:

$$nu_{w,i} + f_i + X_i = 0 \tag{4.4}$$

where $X_x = X_y = 0$ and $X_z = n\rho_w g$ (see Chapter 2).
Let:

$$H = \frac{u_w}{\rho_w g} + z \tag{4.5}$$

where u_w is the pore water pressure, z is the position head and H is the piezometer head or total head.

Substituting Equation (4.3) into Equation (4.4), we can obtain that:

$$n\rho_w g H_{,i} + a_w v_i = 0 \tag{4.6}$$

And let:

$$k = \frac{n\rho_w g}{a_w} \tag{4.7}$$

$$i_i = H_{,i} \tag{4.8}$$

in which k is the coefficient of permeability of the soil in saturated state and i is hydraulic gradient or water potential gradient.

Consequently, Equation (4.6) can be rewritten as:

$$v_i = -ki_i \tag{4.9}$$

Equation (4.9) is the motion equation for seepage in saturated soils. It reflects the correlation between the seepage velocity of pore water and soil-water potential gradient, namely, the velocity of pore water in seepage is proportional to water potential gradient. Darcy obtained this equation via experiment in 1852, and it was known as the famous Darcy's law.

Darcy conducted the seepage test on sands with a vertical tube apparatus schematically shown in Figure 4.1. Test results illustrated that the seepage flux in sand is proportional to the

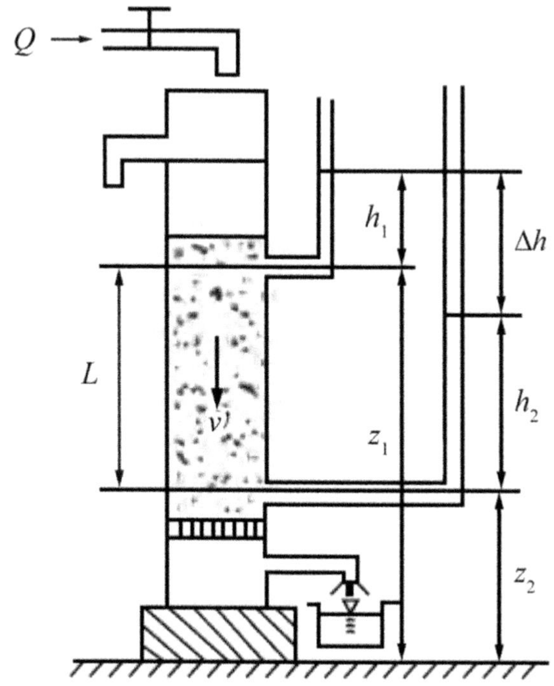

Figure 4.1 The apparatus of Darcy's experiment

sectional area of the tube (A) and the water head loss and inversely proportional to the length of penetration route (L). The results can be expressed as the following equation by introducing the coefficient of permeability k:

$$q = kA\frac{h_2 - h_1}{L} \tag{4.10}$$

or:

$$v = \frac{q}{A} = -ki \tag{4.11}$$

where v is sectional average velocity, $\boldsymbol{i} = (h_2 - h_1)/L$ is the water potential gradient, also called hydraulic gradient, and k is the coefficient of permeability or saturated hydraulic conductivity. It is the seepage velocity on the condition of unit water potential gradient and has the same unit as velocity.

A great amount of subsequent tests showed that the correlation between seepage velocity and hydraulic gradient for other soils such as silt and clay is also consistent with Equation (4.11). This equation is called Darcy's law, which is an experimental law. It illustrates the physical relationship between the moving velocity of pore water and hydraulic gradient.

Equation (4.11) was obtained with homogeneous soils in the invariant flowing status of water. For inhomogeneous soils or non-invariant flowing status, piezometer head varies nonlinearly along the seeping route. Therefore, Darcy's law should be expressed in differential feature, as:

$$v = -k\frac{dH}{dL} \tag{4.12}$$

where the negative sign indicates that the flowing direction of water is opposite to the hydraulic gradient. For the flowing in two-dimensional or three-dimensional spaces, Darcy's law can be written as

$$v_i = -ki_i \tag{4.13}$$

Note that Equation (4.13) is actually Equation (4.9). This means Darcy's law can be derived from the equilibrium differential equation of pore water in the assumption that seepage resistance is proportional to flow velocity. This also means that Darcy's experiments verified the proportional relationship between the force of seepage resistance and flow velocity.

4.1.2 Seepage equation for saturated soils

4.1.2.1 Mass conservation conditions for pore water seepage

In Section 4.1.1, the motion equation that the flowing of pore water should satisfy has been derived based on the equilibrium differential equation of pore water. This motion equation indicates the relationship between seepage velocity and soil-water potential, also called the momentum equation. Only applying the motion equation cannot solve the seepage problems of saturated soils. It also needs to apply the mass conservation condition for pore water, which is the continuous equation of pore water's motion. Combination of the motion equation (equilibrium differential equation) and continuous equation leads to the seepage equation for pore water.

Mass conservation is the theorem that matters commonly obey for moving and variation. For the seepage of pore water, mass conservation means that the difference between the fluxes in and out of a *REV* in soil is equal to the mass change of the water within the *REV* during seepage. When the change of compressibility and density of water can be ignored, the mass conservation condition is the continuous equation of pore water's motion.

Take a free body of *REV* centered with point (x, y, z) as shown in Figure 4.2. The lengths of the free body in x, y and z direction are Δx, Δy and Δz, respectively. The mass conservation of pore water in the free body is investigated in the time period from t to $t + \Delta t$. The flow velocity components of the pore water in the center of the free body at these three directions are assumed to be v_x, v_y and v_z, and the water density is ρ_w. Two sides, $ABCD$ and $A'B'C'D'$, parallel to the coordinate plane yoz are taken, of which their areas are $\Delta y\Delta z$. The water flux at unit area flowing from the left bonding plane $ABCD$ is $v_x - \frac{1}{2}\frac{\partial v_x}{\partial x}\Delta x$, and after the time instant Δt the water mass flowing into the free body from this plane is:

$$\rho_w v_x \Delta y \Delta z \Delta t - \frac{1}{2}\frac{\partial(\rho_w v_x)}{\partial x}\Delta x \Delta y \Delta z \Delta t \tag{4.14}$$

The water flux flowing out of the free body from the right boundary $A'B'C'D'$ is $v_x + \frac{1}{2}\frac{\partial v_x}{\partial x}\Delta x$; after the time instant Δt the water mass flowing out of the free body from this boundary is:

$$\rho_w v_x \Delta y \Delta z \Delta t + \frac{1}{2}\frac{\partial(\rho_w v_x)}{\partial x}\Delta x \Delta y \Delta z \Delta t \tag{4.15}$$

Therefore, the mass difference between the pore water flowing in and out of the free body along the x direction is:

$$-\frac{\partial(\rho_w v_x)}{\partial x}\Delta x \Delta y \Delta z \Delta t \tag{4.16}$$

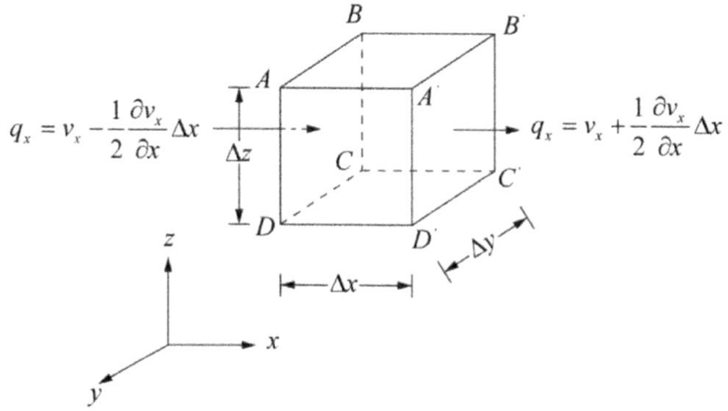

Figure 4.2 The free body of soil with mass change of pore water

Similarly, the mass differences between the pore water flowing in and out of the free body along the y and z directions can be derived as $-\frac{\partial(\rho_w v_y)}{\partial y}\Delta x\Delta y\Delta z\Delta t$ and $-\frac{\partial(\rho_w v_z)}{\partial z}\Delta x\Delta y\Delta z\Delta t$, respectively.

In this case, in the time period of Δt, the total mass difference between the pore water flowing in and out of the free body is:

$$-\left[\frac{\partial(\rho_w v_x)}{\partial x}+\frac{\partial(\rho_w v_y)}{\partial y}+\frac{\partial(\rho_w v_z)}{\partial z}\right]\Delta x\Delta y\Delta z\Delta t \tag{4.17}$$

For saturated soils, the pore water mass in the free body is $\rho_w n\Delta x\Delta y\Delta z$, or $\rho_w nV$. The porosity is n, corresponding to the volumetric water content, and V is the volume of the free body. When the density of pore water is kept invariant, the mass density of water ρ_w is a constant. The mass change of pore water ΔW_w can be obtained by taking full increments on $(\rho_w nV)$, as:

$$\Delta w_w = \rho_w(\Delta nV + n\Delta V) \tag{4.18}$$

When the volumetric deformation of skeleton particles themselves is neglected, $\Delta n = (1-n)\Delta\varepsilon_V$, where $\Delta\varepsilon_V = \frac{\Delta V_v}{V} = \frac{\Delta V}{V}$ is the volumetric strain of soil. Therefore, the mass change of the pore water in the unit volume of the free body in the time period Δt is $\frac{\partial\varepsilon_V}{\partial t}$.

The mass change of pore water in the free body results from the mass difference of the pore water flowing in and out of the free body. Based on the mass conservation theorem, the mass change and mass difference are numerically the same. When pore water is incompressible, i.e., water density ρ_w is a constant, the continuous equation of pore water seepage in saturated soils can be obtained, which is the mass conservation condition, as follows:

$$\frac{\partial\varepsilon_V}{\partial t} = -\left[\frac{\partial v_x}{\partial x}+\frac{\partial v_y}{\partial y}+\frac{\partial v_z}{\partial z}\right] \tag{4.19}$$

where ε_V is the volumetric strain of soils. Equation (4.19) can be expressed as:

$$\frac{\partial\varepsilon_V}{\partial t} = -\nabla\cdot v \tag{4.20}$$

where $\nabla\cdot v$, or $div\,\vec{v}$ is the divergence of v.

4.1.2.2 Seepage equation of saturated soils

Combining the motion equation and the continuous equation of saturated soils leads to the seepage equation. Specifically, substituting the relationship between seepage velocity and soil-water potential gradient expressed by Equation (4.9) into the continuous Equation (4.20), we can derive the basic equation of laminar flow seepage for saturated soils:

$$\frac{\partial \varepsilon_V}{\partial t} = -\nabla \cdot [k \nabla H] \tag{4.21}$$

which can be expanded as:

$$\frac{\partial \varepsilon_V}{\partial t} = \frac{\partial}{\partial x}\left[k_x \frac{\partial H}{\partial x}\right] + \frac{\partial}{\partial y}\left[k_y \frac{\partial H}{\partial y}\right] + \frac{\partial}{\partial z}\left[k_z \frac{\partial H}{\partial z}\right] \tag{4.22}$$

For homogeneous soils, $k_x = k_y = k_x = k$, and if the effect of the volumetric deformation of soil can be ignored, Equation (4.22) is simplified as:

$$\frac{\partial}{\partial x}\left[\frac{\partial H}{\partial x}\right] + \frac{\partial}{\partial y}\left[\frac{\partial H}{\partial y}\right] + \frac{\partial}{\partial z}\left[\frac{\partial H}{\partial z}\right] = 0 \tag{4.23}$$

4.2 Seepage equation for unsaturated soils

Similar to saturated soils, the seepage equation for unsaturated soils also can be derived by combining the motion equation and continuous equation. Deriving the motion equation of pore water needs to apply the equilibrium equation, which contains term of soil-water interaction force. Therefore, we need to analyze the soil-water interaction force for unsaturated soils primarily.

4.2.1 Soil-water interaction force of unsaturated soils

According to the force analysis of the free body of the soil skeleton and pore water in Chapter 2, the soil-water interaction force in saturated soils only includes seepage force, while there is no interaction force between the soil skeleton and pore water when seepage does not exist; for unsaturated soils, except the forces induced by seepage (i.e., the force induced by the relative motion between pore water and the soil skeleton), the nonuniform distribution of pore water will also induce the interaction force between the soil skeleton and pore water, even though there is no water motion.

As shown in Figure 4.3, assume that the volumetric water content distribution in the soil layers above the ground water level in static equilibrium condition is:

$$n_{w0(z_0)} = nS_{0(z_0)} \tag{4.24}$$

where $n_{w0(z_0)}$ is the function of porosity corresponded to the pore water (volumetric water content function) along the vertical direction in static equilibrium condition, n is the porosity, $nS_{0(z_0)}$ is the function of the saturation degree along vertical direction in static equilibrium condition and z_0 is the vertical height from the ground water level.

In Section 1.6, we analyzed the pressure distribution of static pore water in the unsaturated and saturated soil layers above ground water level and understood that pore water pressure presents a continuously linear distribution along the height above and below ground

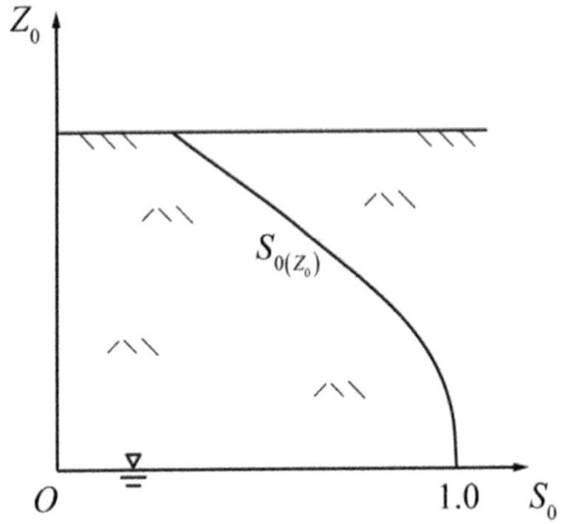

Figure 4.3 Water content distribution of the soil layer above ground water level

water level. Substituting the equation of pore water pressure distribution, $u_w = \psi_m = -\rho_w g z_0$, into the equilibrium differential equation of pore water, the soil-water interaction force in the unit volume of the soil body in saturated and unsaturated soil layers under static equilibrium conditions can be obtained. In saturated soil, the interaction force between the soil skeleton and pore water is equal to zero; while in unsaturated soils:

$$f_{sw0x} = n\rho_w g \frac{\partial S_0}{\partial x} z_0, f_{sw0z} = n\rho_w g \frac{\partial S_0}{\partial z} z_0 \quad (4.25)$$

The components of soil-water interaction force in the X and Z axis are denoted with f_{sw0x} and f_{sw0z}. Equation (4.25) illustrates that the interaction force between the soil skeleton and pore water in the unit volume of soil mass is equal to the variation ratio of pore water weight of the water column at the unit area above ground water level when the pore water is in static equilibrium state. This ratio is introduced by the change of water content, which indicates the variation of the suction force on pore water by soils. If the suction force in the free body varies, the soil-water interaction force is nonzero. Taking the vertical direction, for example, the curve in Figure 4.3 represents the water content distribution in unsaturated soil layers under static equilibrium status, which is the soil-water characteristic curve. The variation of water content along elevation reflects that of the suction on pore water by the soil skeleton, which can be obtained with the corresponding weight of water column. When the volumetric water content changes from nS_0 to $n(S_0 + \frac{\partial S_0}{\partial z} dS_0)$, the weight of the suction water of the unit volume of soil mass increases by $\Delta w = \rho_w gzn\Delta S_0$. Consequently, we have $f_{sw0z} = n\rho_w g \frac{\partial S_0}{\partial z} z_0$.

For the unsaturated soil body isolated from underground soil layers, e.g., soil specimen, how do we calculate the interaction force between the soil skeleton and pore water that is in static equilibrium state? As mentioned previously, for the unsaturated soil layers above the ground water level, the distribution of water content along the elevation matches the water characteristic curve when the pore water is in static equilibrium status. The value of z_0 in the equation is the height from the ground water level to the investigated point. Therefore, for the soil specimen with any water content in which the pore water is in static equilibrium state, we can determine z_0 following the water content based on the soil-water characteristic curve. In other words, we can back-calculate the height of soil layers above the ground water level with the matric potential (suction) in the soil-water characteristic curve corresponding to the water content, as shown in Figure 4.4.

On the other hand, if soil skeleton water is treated as the soil skeleton, it will not participate in seepage. So the effective water content and effective degree of saturation are used to represent the water content and degree of saturation for the soils, respectively. Therefore, Equation (4.24) can be written as:

$$n_{ew0(z_0)} = n_e S_{e0(z_0)} \tag{4.26}$$

where $n_{ew0(z_0)}$ represents the variation of the porosity corresponding to the pore water in which the soil skeleton water is not contained (i.e., effective volumetric water content), along the vertical direction under static equilibrium conditions; n_e is the effective porosity, $n_e = n - n_r$, which is the porosity occupied by the pore water excluding soil skeleton water; and $S_{e0(z_0)}$ represents the effective degree of saturation along the vertical direction in static equilibrium conditions. It should be clarified that, for homogeneous soil mass, the residual water content does not vary in the horizontal or vertical direction, while the variation rate of absolute water content is the same as that of effective water content. Therefore:

$$f_{sw0x} = n_e \rho_w g \frac{\partial S_{e0}}{\partial x} z_0, f_{sw0z} = n_e \rho_w g \frac{\partial S_{e0}}{\partial z} z_0 \tag{4.27}$$

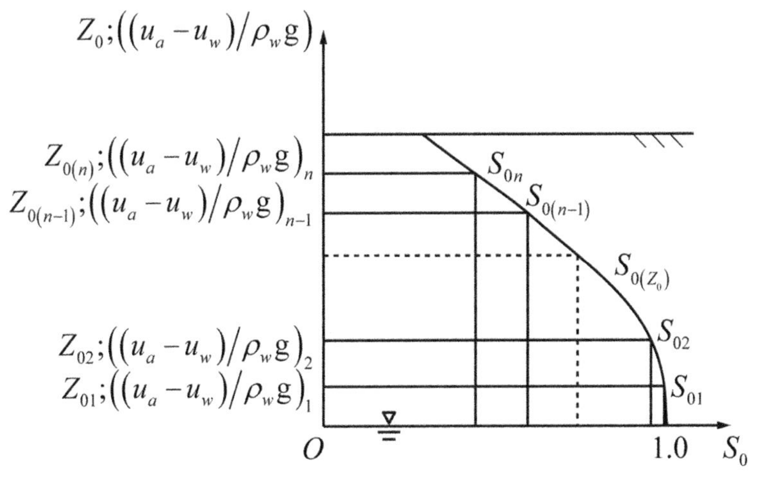

Figure 4.4 Soil-water characteristic curve, water content distribution along the elevation and soil-water interaction force

When the pore water in soil is moving, the distribution of water content does not follow the soil-water characteristic curve. In this case, we still use Equation (4.27) to calculate the soil-water interaction force induced by the nonuniform distribution of pore water, i.e., substituting S_{e0} in the equation by S_e, and then:

$$f_{sw0x} = n_e \rho_w g \frac{\partial S_e}{\partial x} z_0, f_{sw0z} = n_e \rho_w g \frac{\partial S_e}{\partial z} z_0 \tag{4.28}$$

When the seepage of pore water appears in unsaturated soils, we still assume that the resistance force on pore water's motion is proportional to the velocity of pore water. In addition, the seepage resistance force on the pore water in saturated soils and that in unsaturated soils are the same, as far as the water in these two states has the same volume. Therefore, the seepage resistance force on the pore water in the unit volume of unsaturated soil is:

$$\vec{f}_{swf} = a_w \vec{v} \tag{4.29}$$

where a_w is the seepage force factor of the unit volume of pore water (rather than the pore water in the unit volume of soil mass), $\vec{v}$ is the velocity of pore water, and $\vec{f}_{swf}$ is the soil-water interaction force due to seepage. The coordinate components of $\vec{f}_{swf}$ are:

$$f_{swfx} = a_w v_x, f_{swfz} = a_w v_z \tag{4.30}$$

where v_x and v_z are the components of seepage velocity of pore water.

The soil-water interaction force in the equilibrium equation of unsaturated soils includes the forces due to pore water's motion and the nonuniform distribution of pore water, which can be expressed as:

$$\vec{f}_{sw} = \vec{f}_{swf} + \vec{f}_{sw0} \tag{4.31}$$

where a_w is the soil-water interaction factor of the unit volume of pore water in unsaturated soils and $\vec{f}_{sw0}$ represents the soil-water interaction due to the nonuniform distribution of pore water, of which the components were given in Equation (4.32). Consequently, the total soil-water interaction forces in x and z directions are:

$$f_{swx} = a_x v_x + \rho_w g n_e \frac{\partial S_e}{\partial x} z_0 \tag{4.32}$$

$$f_{swz} = a_z v_z + \rho_w g n_e \frac{\partial S_e}{\partial z} z_0 \tag{4.33}$$

4.2.2 Motion equation and the coefficient of permeability for unsaturated soils

For the seepage in unsaturated soils, the motion equation is the formula of Darcy's law. Different from saturated soils, for unsaturated soils, the velocity of pore water is not proportional to the soil-water potential gradient. Therefore, the formula of Darcy's law for unsaturated soils is only a formal expression.

As previously mentioned, the total water head of the pore water in unsaturated soils consists of gravity potential Ψ_g and matrix potential Ψ_m. The soil-water potential expressed with the potential energy of the unit weight of pore water can be written as:

$$H = z + \psi_m \tag{4.34}$$

or

$$H = z + \frac{u_w}{\rho_w g} \tag{4.35}$$

where H is the total head, z is the gravity head, u_w is the pore water pressure of unsaturated soils, which is the matric potential of the unit volume of pore water, ρ_w is the density of water and g is the gravitational acceleration.

Let us ignore the interaction force between the pore water and pore air and introduce the equilibrium differential equation of pore water in unsaturated soils, which are:

$$\frac{\partial}{\partial x}(n_{ew}u_w) + f_{swx} = 0 \tag{4.36}$$

$$\frac{\partial}{\partial z}(n_{ew}u_w) + f_{swz} + n_{ew}\rho_w g = 0 \tag{4.37}$$

where f_{swx} and f_{swz} are the interaction forces between the pore water and the soil skeleton in x and z direction, respectively; and n_{ew} is the porosity corresponding to the pore water excluding soil skeleton water, which is the effective volumetric water content.

The effective degree of saturation is

$$S_e = \frac{n_{ew}}{n_e} = \frac{S - S_r}{1 - S_r} \tag{4.38}$$

The velocity components of pore water are v_x and v_z. Substituting the expression of soil-water interaction forces for unsaturated soils, Equations (4.32) and (4.33), into the equilibrium differential equation, Equations (4.36) and (4.37), respectively, can obtain the following expressions after reorganization:

$$v_x = -k_x\left[\frac{\partial(S_e H)}{\partial x} - (z - z_0)\frac{\partial S_e}{\partial x}\right] \tag{4.39}$$

$$v_z = -k_z\left[\frac{\partial(S_e H)}{\partial z} - (z - z_0)\frac{\partial S_e}{\partial z}\right] \tag{4.40}$$

where $k_x = \frac{n_e\rho_w g}{a_{wx}}$ and $k_z = \frac{n_e\rho_w g}{a_{wz}}$ are supposed to be the permeability coefficient in x and z directions, like that of saturated soils, S_e is the effective degree of saturation, and H is the total head.

Furthermore, Equations (4.39) and (4.40) can be written as:

$$v_x = -k_{ux} \frac{\partial H}{\partial x} \tag{4.41}$$

$$v_z = -k_{uz} \frac{\partial H}{\partial z} \tag{4.42}$$

where k_{ux} and k_{uz} are the permeability coefficient functions for unsaturated soils in x and z direction, and:

$$k_u = k\left[S_e + (H - z + z_0)\frac{\partial S_e}{\partial H}\right] \tag{4.43}$$

where k has the same expression and physical meaning as the permeability coefficient for saturated soils. That said, if the respective resistance force due to motion on the unit volume of pore water is the same for saturated and unsaturated soils, the factor k in Equation (4.43) is the same as the permeability coefficient for saturated soils. The relationship between the permeability coefficient for unsaturated soils and that for saturated soils is therefore established by Equation (4.43). Apparently, besides its correlation to the permeability coefficient for saturated soils, the permeability coefficient for unsaturated soils is also related to the water content and soil-water potential.

4.2.3 Seepage equation for unsaturated soils

4.2.3.1 Continuous equation of the seepage in unsaturated soils

Besides the previously derived motion equation for pore water in unsaturated soils, the continuous equation is also needed in order to obtain the seepage equation. Similar to the case of saturated soils, a free body is taken as shown in Figure 4.2. Based on the mass conservation condition of the pore water in the free body, the continuous equation of unsaturated soils without counting the volumetric deformation of soil can be obtained:

$$\frac{\partial n_{ew}}{\partial t} = -\left[\frac{\partial v_x}{\partial x} + \frac{\partial v_y}{\partial y} + \frac{\partial v_z}{\partial z}\right] \tag{4.44}$$

or:

$$\frac{\partial n_{ew}}{\partial t} = -\nabla \cdot v \tag{4.45}$$

where n_{ew} is the effective volumetric water content of soil.

When the volumetric deformation of soil is considered, the change of the mass of pore water in the free body is the summation of two parts: (1) that due to the volumetric change of the free body and (2) that due to the change of water content. The relationship can be expressed by the formula:

$$\rho_w n_{ew} \Delta\varepsilon_V \Delta x \Delta y \Delta z + \rho_w \Delta n_{ew} \Delta x \Delta y \Delta z \tag{4.46}$$

Substituting$n_{ew} = n_e S_e$, $\Delta n_{ew} = \Delta(n_e S_e) = n_e \Delta S_e + S_e \Delta n_e$ and $\Delta n_e = (1-n_e)\Delta\varepsilon_V$ into equation (4.46), the change of the mass of the pore water in the free body is:

$$\begin{aligned}&\rho_w n_{ew} \Delta\varepsilon_V \Delta x \Delta y \Delta z + \rho_w \Delta n_{ew} \Delta x \Delta y \Delta z\\ &= \rho_w n_e \cdot S_e \Delta\varepsilon_V \Delta x \Delta y \Delta z + \rho_w [n_e \cdot \Delta S_e + S_e (1-n_e)\Delta\varepsilon_V] \Delta x \Delta y \Delta z\\ &= \rho_w (n_e \cdot \Delta S_e + S_e \cdot \Delta\varepsilon_V) \Delta x \Delta y \Delta z\end{aligned} \tag{4.47}$$

The change rate with time of the mass of pore water in a unit volume of soil is:

$$n_e \frac{\partial(\rho_w S_e)}{\partial t} + S_e \frac{\partial(\rho_w \varepsilon_V)}{\partial t} \tag{4.48}$$

In fact, the pore water mass corresponding to the effective water content is $\rho_w n_{ew} = n_e \rho_w S_e$. Consequently, the total difference of time is Equation (4.48).

It is known, according to the mass conservation theorem, that the mass change of the pore water in the free body is the mass difference of the pore water flowing in and out of the free body. Therefore, the continuous equation for the motion of pore water is:

$$n_e \frac{\partial(\rho_w S_e)}{\partial t} + S_e \frac{\partial(\rho_w \varepsilon_V)}{\partial t} = -[\frac{\partial(\rho_w v_x)}{\partial x} + \frac{\partial(\rho_w v_y)}{\partial y} + \frac{\partial(\rho_w v_z)}{\partial z}] \tag{4.49}$$

When the water body is incompressible and the density of water ρ_w is constant, Equation (4.49) can be written as:

$$n_e \frac{\partial S_e}{\partial t} + S_e \frac{\partial \varepsilon_V}{\partial t} = -(\frac{\partial v_x}{\partial x} + \frac{\partial v_y}{\partial y} + \frac{\partial v_z}{\partial z}) \tag{4.50}$$

When the soil body is saturated, S_e is equal to 1 and Equation (4.50) is simplified as:

$$\frac{\partial \varepsilon_V}{\partial t} = -(\frac{\partial v_x}{\partial x} + \frac{\partial v_y}{\partial y} + \frac{\partial v_z}{\partial z}) \tag{4.51}$$

which is the same as the continuous equation of the pore water seepage in saturated soils, Equation (4.19).

4.2.3.2 Seepage equation for unsaturated soils

Similarly, the seepage equation for unsaturated soils can be obtained by combining the motion equation and continuous equation. Specifically, expanding the motion equation of seepage, Equations (4.39) and (4.40), to a three-dimensional situation and substituting them into the continuous equation (4.50), the seepage equation for unsaturated soils can be obtained:

$$n_e \frac{\partial S_e}{\partial t} + S_e \frac{\partial \varepsilon_V}{\partial t} = -\nabla \cdot [k_u \nabla H] \tag{4.52}$$

where the permeability coefficient k_u is expressed with Equation (4.43).

4.3 Formula of seepage force

In soil mechanics, the seepage force is generally defined as the interaction force on the soil skeleton by pore water in the unit volume of soil mass. Seepage force is a type of body force with the same dimension as the volume-weight of water and the application direction consistent with the seepage velocity. Since the seepage force on the soil skeleton has the same order of magnitude as and the opposite direction from the seepage resistance on the pore water applied by the soil skeleton, the expression of seepage force can be derived by analyzing the force on pore water.

The expression of seepage force of saturated soil given in almost all the soil mechanics textbooks is:

$$\boldsymbol{J} = \gamma_w \boldsymbol{i} \tag{4.53}$$

where $\boldsymbol{J}$ represents the seepage force (vector), γ_w is the volume-weight of water and $\boldsymbol{i}$ is the soil-water potential gradient (vector).

According to the deduction procedure of the equilibrium differential equation of the soil skeleton and pore water, the seepage force defined in Equation (4.53) is the same as the interaction force between the soil skeleton and pore water in the equilibrium equation of pore water in saturated soils (as shown in Chapter 2). The interaction force between the soil skeleton and pore water in the unit volume of soil mass obtained via the equilibrium differential equation of pore water (Equation (2.10)) is:

$$\boldsymbol{f} = n\gamma_w \boldsymbol{i} \tag{4.54}$$

where n is the porosity of soils, γ_w is the unit weight of water, and $\boldsymbol{i}$ is the water potential gradient (vector).

As $\boldsymbol{J}$ and $\boldsymbol{f}$ represents the same meaning, why is there a factor n in the expression of $\boldsymbol{f}$ but not in that of $\boldsymbol{J}$? This is due to the derivation procedure of $\boldsymbol{J}$ in soil mechanics textbooks, which implied that the definition of $\boldsymbol{J}$ is the seepage force on the unit volume of pore water, not on the pore water in the unit volume of soil mass.

In the textbooks of soil mechanics, the expression of seepage force is generally obtained by conducting equilibrium analysis on the free body of pore water in a vertical direction or diagonal direction. This can be illustrated by taking the relevant description in *Advanced Soil Mechanics* by Braja (2008) as an example:

> *To evaluate the seepage force per unit volume of soil, consider a soil mass bounded by two flow lines ab and cd and two equipotential lines ef and gh, as shown in Figure 4.5. The soil mass has unit thickness at right angles to the section shown. The self-weight of the soil mass is (length)(width)(thickness)(γ_{sat}) = (L)(L)(1)(γ_{sat}) = $L^2\gamma_{sat}$. The hydrostatic force on the side ef of the soil mass is (pressure head)(L)(1) = $h_1\gamma_w L$. The hydrostatic force on the side gh of the soil mass is $h_2\gamma_w L$. For equilibrium,*

$$\Delta F = h_1\gamma_w L + L^2\gamma_{sat}\sin\alpha - h_2\gamma_w L \tag{A1}$$

However, $h_1 + L\sin\alpha = h_2 + \Delta h$, *so*

$$h_2 = h_1 + L\sin\alpha - \Delta h \tag{A2}$$

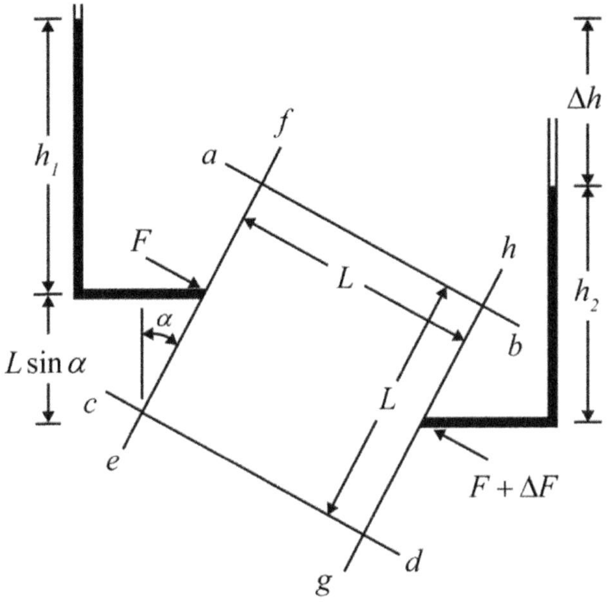

Figure 4.5 Seepage force determination (Braja, 2008)

Combining equations (A1) and (A2),

$$\Delta F = h_1\gamma_w L + L^2\gamma_{sat}\sin\alpha - (h_1 + L\sin\alpha - \Delta h)\gamma_w L \tag{A3}$$

$$\Delta F = L^2(\gamma_{sat} - \gamma_w)\sin\alpha + \Delta h\gamma_w L = L^2\gamma'\sin\alpha + \Delta h\gamma_w L \tag{A4}$$

$L^2\gamma'\sin\alpha$ is the submerged unit weight of soil, and $\Delta h\gamma_w L$ is the seepage force, where $\gamma' = \gamma_{sat} - \gamma_w$. From Eq. (A4) we can see that the seepage force on the soil mass considered is equal to $\Delta h\gamma_w L$. Therefore,

$$\boldsymbol{J} = \frac{\Delta h\gamma_w L}{L^2} = \gamma_w\frac{\Delta h}{L} = \gamma_w i \tag{A5}$$

where i is the hydraulic gradient.

Now the forces of the free body of the pore water in the soil column are analyzed in order to obtain the formula of seepage force on the definition that the seepage force is the interaction force between skeleton and pore water in the unit soil. For the free body of pore water in the inclined column of soil, as shown in Figure 4.6, the forces on the free body of pore water include:

1. The self-weight of the pore water is (n)(length)(width)(thickness)(γ_w) = $(n)(L)(L)(1)(\gamma_w) = nL^2\gamma_w$;

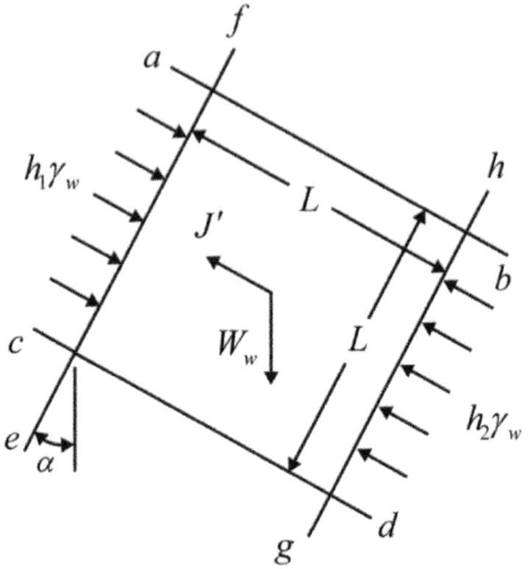

Figure 4.6 The force analysis of the free body of pore water

2. The hydrostatic force on the side *ef* of the pore water body is (pressure head)(*n*) (*L*) (1) = $nh_1L\gamma_w$, and the hydrostatic force on the side *gh* of the pore water body is $nh_2L\gamma_w$ (when the area of the soil column is assumed to be *L*, the area of the column of pore water is *nL*);
3. The resistant force is applied on the flowing water by the soil skeleton, which is equal to seepage force in quantity but opposite in direction. Assuming that the resistant force is j' in unit soil mass, the sum of the resistant force is $J' = j'L^2 = J$.

According to the condition of the equilibrium of the inclined free body of pore water, it can be obtained that:

$$nh_1L\gamma_w + W_w \sin\alpha + J' = nh_2L\gamma_w \tag{4.55}$$

$$nh_1L\gamma_w + nL^2\gamma_w \sin\alpha + j'L^2 = nh_2L\gamma_w \tag{4.56}$$

$$j' = \frac{nL\gamma_w(h_1 + L\sin\alpha - h_2)}{L^2} = \frac{n\gamma_w L\Delta h}{L^2} = n\gamma_w i \tag{4.57}$$

Another example is taken from *Soil Mechanics* written by Chen *et al.* (1994), which is a vertical placed soil column in a stable seepage test, as shown in Figure 4.7. The forces of the free bodies of the soil skeleton and pore water are presented. When the force analysis was respectively conducted on free body of the soil skeleton and pore water that were taken separately, the seepage force is expressed as an external force, as shown in Figure 4.7. The following is the original text in *Soil Mechanics* that describes the force and equilibrium analysis procedure on the free body of pore water.

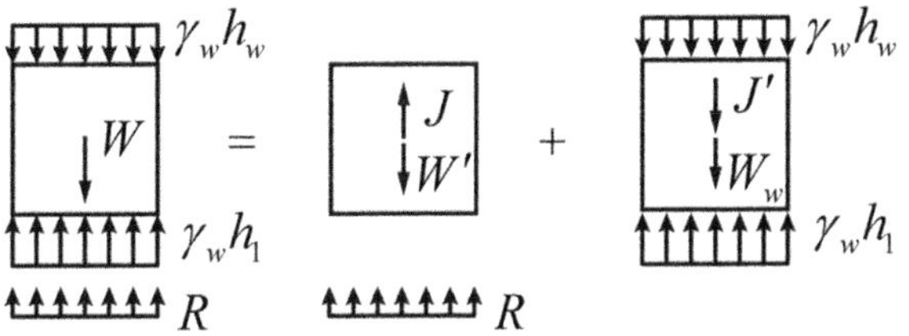

(a) Water-soil entity (b) Soil skeleton (c) Water body

Figure 4.7 Two isolation approaches types of the free body in seepage

The forces applied on the free body of pore water in soil column (as presented in Fig. 4.7(c)) include:

(1) *The sum of the gravity of pore water and the buoyancy of soil particles. The latter should be equal to the gravity of the water with the same volume as the soil particles. This can be expressed as*

$$W_w = V_v\gamma_w + V_s\gamma_w = V\gamma_w = L\gamma_w \tag{B1}$$

It can be known that W_w is the weight of the water column with the length of L.

(2) *The water pressure at the boundary of the top and bottom sides of water column, $\gamma_w h_w$ and $\gamma_w h_1$, and*
(3) *The resistance force on the water flow applied by the soil particles in soil column. It has the same amount but opposite direction as the seepage force. If the resistance force on the water flow applied by the soil particles in the unit of soil body is assumed to be j', the total resistance force is $J' = j'L = J$, of which the direction is vertically downward.*

Considering the equilibrium conditions of the free body of pore water (as shown in Fig. 4.7(c)) can obtain:

$$\gamma_w h_w + W_w + J' = \gamma_w h_1 \tag{B2}$$

$$\gamma_w h_w + L\gamma_w + j'L = \gamma_w h_1 \tag{B3}$$

$$j' = \frac{\gamma_w\left(h_1 - h_w - L\right)}{L} = \frac{\gamma_w \Delta h}{L} = \gamma_w i \tag{B4}$$

Therefore, the seepage force is $j = j' = \gamma_w i$.

It can be observed that the area of the free body of pore water is assumed to be a unit area in this force and equilibrium analysis.

If the seepage force is defined as the force on the soil skeleton applied by pore water within the unit of soil body, i.e., the force on the pore water by the soil skeleton in the unit of soil body (the unit volume of soil rather than the unit volume of pore water), the force analysis on the aforementioned free body of pore water in soil column should be:

(1) The gravity of pore water, which is the weight of the pore water in the soil column: $W_w = \gamma_w V_v = n\gamma_w V = n\gamma_w L$;
(2) The water pressure at the boundaries on the top and bottom sides of water column, $\gamma_w h_w$ and $\gamma_w h_1$ (when the area of the soil column is 1, the area of pore water column is n);
(3) The resistance force on the flowing water applied by the soil skeleton in the soil column, which has the same amount but opposite direction as the seepage force. If the resistance force on the water by the skeleton in the unit of soil body is assumed to be j', the total resistance force is $J' = j'L = J$, of which the direction is vertically downward.

Considering the equilibrium conditions for the free body of water (as shown in Fig. 4.7(c)), we have:

$$n\gamma_w h_w + W_w + J' = n\gamma_w h_1 \tag{4.58}$$

$$n\gamma_w h_w + nL\gamma_w + j'L = n\gamma_w h_1 \tag{4.59}$$

$$j' = \frac{n\gamma_w (h_1 - h_w - L)}{L} = \frac{n\gamma_w \Delta h}{L} = n\gamma_w i \tag{4.60}$$

Therefore, the seepage force is $\boldsymbol{j} = \boldsymbol{j}' = n\gamma_w \boldsymbol{i}$, which is the same as the result derived by applying the equilibrium differential equation of pore water.

Certainly, if the seepage force is defined as the resistance force on a unit volume of pore water by the soil skeleton, the expression of seepage force is:

$$\boldsymbol{j} = \boldsymbol{j}' = \gamma_w \boldsymbol{i} \tag{4.61}$$

where $\boldsymbol{j}$ is the seepage force on a unit volume of water. If $\boldsymbol{f}$ is the seepage force on the pore water in a unit volume of soil, then:

$$\boldsymbol{f} = n\boldsymbol{j} \tag{4.62}$$

This correlation can also be obtained according to the fact that the seepage forces on the same section of soil body are the same. Since $f \cdot A = j \cdot A_w$ and $A_w = n \cdot A$ (A is the sectional area), then $f \cdot A = j \cdot A_w = j \cdot n \cdot A$, i.e., $f = j \cdot n$.

Another question in this discussion is: should the counterforce of the buoyancy of the soil skeleton be included in the gravity (force) of pore water? The answer is NO. The reasons are as follows:

In static equilibrium conditions, the self-weight of pore water induces the variation of pressure; the action of pressure results in the buoyancy of soil skeleton particles, whose value is equal to the weight of the water replaced by the soil skeleton particles. The counterforce to the buoyancy of soil particles is actually water pressure, which does not vary the weight of water nor the pressure of water. No matter how the soil particles interact with pore water, the pore water pressure on the surface of the free body of pore water remains unchanged. This has been illustrated during the deduction procedure of equilibrium differential equations in Chapter 3. Consequently, when the pore water is designated as the analysis subject, its gravity should not include the counterforce of buoyancy force.

The expression of the seepage force for unsaturated soils is analyzed as follows. The seepage in unsaturated soils includes the seepage of pore water and pore air. Both the forces induced by the seepage of pore water and that of pore air are called the seepage force of unsaturated soils. Similar to saturated soil, the seepage force of the pore water in unsaturated soils is defined as the force on the soil skeleton by the pore water in a unit volume of the soil body; the counterpart of the pore air in unsaturated soils is defined as the force on the soil skeleton by the pore air in a unit volume of the soil body. They are equal to the seepage resistance on the pore water and pore air applied by the soil skeleton in a unit volume of the soil body, respectively. According to these definitions, we can use the equilibrium differential equations to derive the seepage force formula of unsaturated soils.

Take the interaction force between pore water and the soil skeleton, for example. Assuming the water content of soils is uniform, the equilibrium differential equation of the pore water in unsaturated soils (Equation (2.11)) is:

$$\begin{cases} n_w \dfrac{\partial u_w}{\partial x} + f_{swx} = 0 \\ n_w \dfrac{\partial u_w}{\partial z} + f_{swz} + n_w \gamma_w = 0 \end{cases} \tag{4.63}$$

When the atmosphere is open and the pore air pressure and solute potential can both be neglected, the expression of total water potential (total water head) is Equation (4.39). Consequently, the seepage force of the unsaturated soils with uniform water content can be obtained based on Equation (4.63), as:

$$\boldsymbol{f} = n_w \gamma_w \boldsymbol{i} \tag{4.64}$$

where n_w is the porosity corresponding to the pore water, i.e., the volumetric water content.

It can be seen that the expression of seepage force for unsaturated soils in the condition of uniform water content is the same as that for saturated soils, where only the n_w is the porosity corresponding to the pore water.

This is the interaction force between pore water and the soil skeleton corresponding to the water potential gradient when the water content is uniform. In general, the water content of unsaturated soils is nonuniform. In this case, it can be found, based on the equilibrium equation, that the interaction force between the soil skeleton and the pore water does not only include the applied force corresponding to water potential gradient but also the force corresponding to the change of water content. The expression of the interaction force is:

$$\begin{cases} f_x = n_w \gamma_w i_x + u_w \dfrac{\partial n_w}{\partial x} \\ f_z = n_w \gamma_w i_z + u_w \dfrac{\partial n_w}{\partial z} \end{cases} \tag{4.65}$$

The interaction force corresponding to the change of water content behaves as capillary action, which is essentially induced by surface tension.

Now the question is briefly discussed herein whether the interaction force corresponding to the change of water content should be included in the seepage force. It can be understood that the seepage force is induced by the percolation of pore fluid in the soil skeleton. In the condition of laminar seepage, the seepage force is proportional to the seepage velocity (i.e., the velocity of the relative motion between the pore fluid and the soil skeleton). There is no seepage force while the relative motion between the pore fluid and the soil skeleton, i.e. the seepage, does not exist. However, it is known based on the equilibrium equation that there is an interaction force induced by the change of water content for unsaturated soils even in the static equilibrium condition, i.e., where seepage does not exist. Therefore, we suggest that the definition of seepage force should not include the interaction force corresponding to the change of water content. In this case, the interaction force would not be equal to the seepage force, which includes the interaction force induced by the change of water content.

4.4 The overflow condition of the gas in soils

4.4.1 The overflow condition of the closed air bubbles

Generally speaking, soil in a natural state is fairly difficult to be completely saturated, thanks to the prevalent air. The air may be dissolved into the pore water or exist in the form of micro bubbles. The air bubbles can be surrounded by pore water or be absorbed on the soil skeleton particles. The state of micro air bubbles might turn to a free state from a being-absorbed state when the soils are strongly disturbed. The air bubbles would leave the particles and move upwards in pore water or form a larger air bubble together with the others. The air bubbles would form a closed boundary in a soil body when they are large enough, which would exceed the size of the *REV*, as shown in Figure 4.8.

The air pressure inside the closed air bubble is relevant to the normal stress (total stress) of the soil outside the air bubble. In the meantime, the radius of the air bubble can also affect the air pressure while the surface tension of the contractile skin itself on the boundary of the

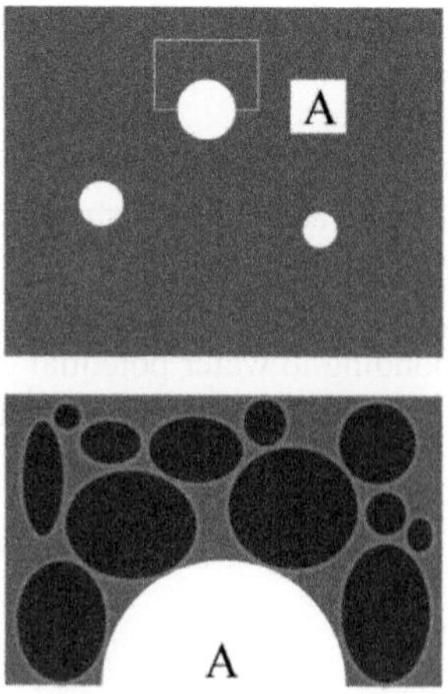

Figure 4.8 The boundary of a closed air bubble

air bubble is considered. With the equilibrium analysis, the air pressure inside the air bubble can be expressed as:

$$u_a = \frac{2T_s}{R} + \sigma_t \tag{4.66}$$

where u_a is the air pressure inside the air bubble, R is the radius of the air bubble and σ_t is the total stress of the soil outside the bubble.

If the pressure difference on the two sides of the contractile skin at some point of the air bubble was larger than the air-entry value at this point of soil, the air inside the air bubble would break through the contractile skin and move upward outside the surface of soil or move to a higher location and reach a new equilibrium state. This situation is called the overflow of air bubble. The overflow condition of air bubble is:

$$u_a - u_w \geq \left(u_a - u_w\right)_b \tag{4.67}$$

where $(u_a - u_w)$ is the pressure difference through the contractile skin and $(u_a - u_w)_b$ is the air-entry value of the saturated soil outside the air bubble.

The air-entry value of soil is the suction of the soil that the outside air has to "break through" when it enters the pores in soil. It is also the maximum air pressure that the saturated soil can "defend" against air entering its saturated pores. This air-entry value is called "displacement pressure" in petroleum engineering and "bubbling pressure" in ceramics engineering. By applying Kelvin's equation, the air-entry value $(u_a - u_w)_b$ can be expressed as:

$$(u_a - u_w)_b = \frac{2T_s}{R_b} \tag{4.68}$$

in which R_b is the maximum radius of the pores in the soil body.

Consequently, the overflow condition of the closed air bubble can be written as:

$$\sigma_t + \frac{2T_s}{R} \geq (u_a - u_w)_b \tag{4.69}$$

The experimental observation showed that it would lead the sand structure to be destroyed or even form piping along its break-through route when a closed air bubble overflows in near-saturated loose sand.

4.4.2 The overflow condition for pore air while surface water is infiltrating

When rainfall is heavy and there is surface ponding, the pore air in soil will form an aeration zone. The pore air pressure in the aeration zone changes with the change of the pressure of water in the overlying soil, while the effect of pore air pressure should not be ignored. The soil skeleton stress can be obtained by applying the soil skeleton stress equation of saturated or unsaturated soils, whereas the pore air pressure is unequal to the atmospheric pressure.

When water on the ground surface infiltrates, if the permeability of soil is larger than the infiltration rate of the ground water, a portion of pore air in soils would be replaced by pore water and freely discharged out of ground surface with the infiltration of surface water. In this case, the overflow condition for the pore air in soil is that the pressure gradient of the pore air at its rising direction is equal to the moving resistance on it, as:

$$\frac{\partial(n_a u_a)}{\partial z} = -f_{saz} \tag{4.70}$$

where f_{saz} is the interaction force between pore air and soils in z direction.

By contrast, if the permeability of soil is far lower than the infiltration rate of the ground water, the ground surface would be covered by water, while the route by which pore air escapes would be blocked. During a short time that the ground was covered by water, the pore air would be compressed. If soil is ideal homogeneous, the infiltration front would stay at the same height. When the pressure of the compressed pore air is equal to the pressure of the pore water at the infiltration front, infiltration will stop, as illustrated in Figure 4.9. However, soils are actually nonideal homogenous, and the infiltration is affected by boundary conditions, which barely ensure the percolating pore water will keep the same velocity at the infiltration front. Therefore, the actual infiltration front is normally uneven, as shown in Figure 4.10. The

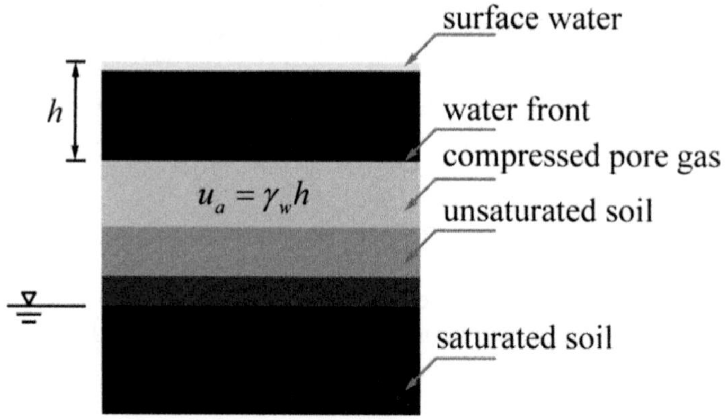

Figure 4.9 The ideal infiltration front at the same level

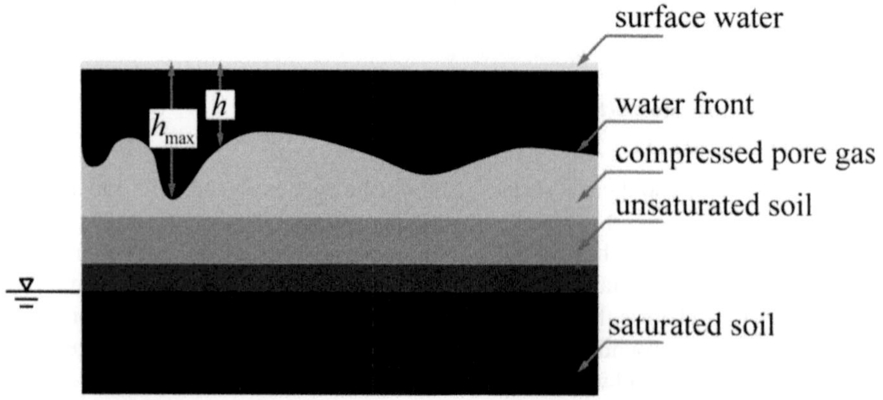

Figure 4.10 The infiltration front and overflow of compressed pore air

deeper the location of infiltration front, the higher the pore water pressure at this point would be. However, all the pressures of the pore air in the area covered with ground water are the same, equal to the maximum water pressure at the infiltration front, i.e., the water pressure at the deepest point of the infiltration front. Therefore, the overflow condition of pore air should satisfy the following terms (a) or (b):

a. the pressure difference at the boundary of infiltration front is higher than the air-entry value of the soil, as:

$$u_a - u_w \geq (u_a - u_w)_b \tag{4.71}$$

or:

$$(h_{\max} - h) \geq \frac{(u_a - u_w)_b}{\gamma_w} \tag{4.72}$$

b. the pore air pressure at the investigated point is higher than the soil total stress, i.e.:

$$\gamma_w h_{\max} \geq \gamma_{sat} h \tag{4.73}$$

or:

$$\frac{h_{\max}}{h} \geq \frac{\gamma_{sat}}{\gamma_w} \tag{4.74}$$

where u_a is the pressure of the compressed pore air, u_w is the pore water pressure at the infiltration front, $(u_a - u_w)_b$ is the air-entry value of the soil at the infiltration front, h_{max} is the maximum percolation depth at the infiltration front, h is the percolation depth at the investigated point and γ_{sat} is the saturated unit weight of the soil.

Once either of the those two conditions, (a) or (b), is satisfied during infiltration, pore air will overflow from the soil. When the pressure difference at the boundary of infiltration front is higher than the air-entry value of the soil, there will be an escaping route for the pore air; while the pore air pressure is higher than the total stress of the overburden soil, a fracture plane or air cushion will form.

When the air in the aeration zone starts to enter the soil layer above the infiltration front, the pore air pressure in the aeration zone decreases and the infiltration accelerates. On the other hand, the higher the water pressure at a point on the infiltration front, the higher the infiltration speed of the water at this point will be, which will also accelerate the overflow of the pore air.

Chapter 5

Discussion on some issues related to effective stress

It is known based on the discussion in Chapter 3 that Terzaghi's effective stress for saturated soils (Terzaghi, 1925, 1936) is actually the soil skeleton stress induced by external forces. This is also called soil skeleton stress due to external forces, and it excludes the effect of pore water pressure. The pore water pressure only induces volumetric change of the soil particles. In addition, only the stress due to pore water pressure on the interface between soil particles contributes to the shear strength of the soil. Since the volumetric change of the soil particle and the shear strength on the interface between soil particles can be ignored in most cases, the strength and deformation of soil are determined by the effective stress. From this perspective, we not only have interpreted Terzaghi's effective stress equation in a novel way but also gave an explicit physical explanation of effective stress.

Then why is effective stress defined as the soil skeleton stress due to external forces? Actually, this question has been answered in Chapter 3. The effect of pore fluid pressure and the other external forces can be revealed when the soil skeleton is taken as the research object for internal force analysis. Uniform pore fluid pressure only causes volumetric change of the skeleton particles, without resulting in shear stress. Only the stress at the interface between soil particles, which is induced by uniform pore fluid pressure, contributes to the shear strength of soils. The skeleton stress induced by the other external forces determines the strength and deformation of soils, when the impact of pore fluid pressure is negligible. Therefore, effective stress can be defined as the soil skeleton stress due to external forces because: (1) the strength and deformation of soils are actually the strength and deformation of the soil skeleton, respectively; (2) the strength and deformation of the soil skeleton are both controlled by its stress; (3) the effect of uniform pore fluid pressure, different from that of the other external forces, only results in volumetric change of soil particles, and only the stress on the interface between soil particles contributes to the shear strength of soils; and (4) the influence of uniform pore fluid pressure on the strength and deformation of soils can be ignored in most cases.

It is well known that the current theory of soil mechanics was built on the concept of effective stress. Therefore, it is convenient to define the soil skeleton stress due to external forces as effective stress; certainly, it is also okay to not call the stress "effective stress." Regardless of its designation, soil skeleton stress is a measurement for the internal force of the soil skeleton. Furthermore, the internal force of the soil skeleton is real, as well as the skeleton stress.

In fact, the strength and deformation of soils, i.e., the strength and deformation of the soil skeleton, are controlled by the soil skeleton stress, which includes two parts, due to pore fluid pressure and effective stress, respectively. These two parts cannot be simply superimposed in the determination of strength or deformation, as the impacts of them on strength and deformation are different. The strength and deformation of soils are governed by effective stress, while the contribution of pore fluid pressure-induced soil skeleton stress on them is negligible. It is important to note that the debates regarding the effective stress, the effective stress equation and their applications in the whole framework of soil mechanics still exist to date, and they are currently being investigated in various ways.

5.1 Does Terzaghi's effective stress equation need to be modified?

Among many perspectives that suggest Terzaghi's effective stress equation be modified, the earliest and the most respective one was proposed by Skempton (1961). In the well-known article "Effective stress in soils, concrete and rocks," Skempton interpreted the effective stress through the stress in soil skeleton particles and deduced a modified effective stress equation via the force equilibrium analysis of soil particles. In addition, Skempton provided the expression of the shear strength and volumetric change of soil, in which the effect of pore water pressure was considered. Furthermore, he also presented the modified effective stress equation based on the strength and volumetric change according to the principle that the strength and deformation of soils were controlled by the effective stress. In this article, Skempton also addressed, via the experimental data: (1) Terzaghi's effective stress equation is valid for saturated soils, and it is unnecessary to be modified; (2) the modified effective stress equation deduced with the force and equilibrium of soil particles is inconsistent with the experimental results; and (3) the modified effective stress expression considering the effect of pore water pressure on the shear strength and volumetric change is applicable for rocks and concrete.

Additionally, Skempton proposed the expression of the shear strength and volumetric change of soils (porous medium), which considered the impact of pore water pressure, i.e., Equations (3.8) and (3.9), and pointed out that the shear strength and volumetric change of soils (porous medium) were not only related to the effective stress but also relevant to the pore water pressure. Unfortunately, by following the principle of "effective stress governs strength and deformation," Skempton's objective in providing Equations (3.8) and (3.9) was to deduce a modified shear strength-equivalent and volumetric change-equivalent effective stress equation. This enhanced two perspectives, which are (1) the effective stress should be deduced based on the principle of the shear strength- and volumetric strain-equivalency and (2) the effective stress equation needs to be modified.

More modified effective stress equations have been proposed thereafter. Lade and Boer (1997) organized those equations and simplified them into:

$$\sigma = \sigma_t - \eta u_w \tag{5.1}$$

and categorized and summarized the equations by the correlations of η to the stress-strain properties and shear strength of soils.

Shao (2011a, 2012) claims that Terzaghi's effective stress equation needs to be modified because of the effect of the pore water pressure. It was included in the resulting interparticle stress, and it was not separated from the force analysis of effective stress with interparticle stress, as shown in Figure 5.1. The equilibrium equation of the perpendicular interparticle stress given in many current soil mechanics books is:

$$P = P_s + u_w(A - A_s) \tag{5.2}$$

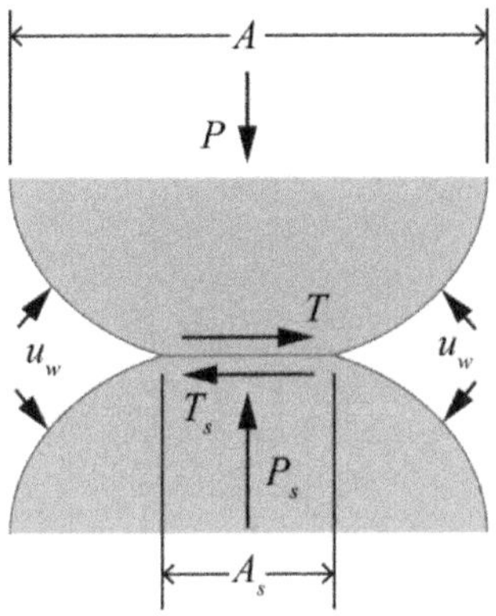

Figure 5.1 Pore water pressure and the contact force between particles

and $\sigma_t = P/A$ is used to denote the total stress, $\sigma' = P_s/A$ to denote the effective stress, and $a_c = A_s/A$ to represent the contact area ratio. Therefore, Equation (5.2) can be written as:

$$\sigma_t = \sigma' + (1 - a_c)u_w \tag{5.3}$$

which is the so-called modified effective stress equation. However, as we know based on the discussion in Chapter 2, the internal force among the particles due to pore water pressure and the pore water pressure applied on soil particle surfaces are equilibrated. By using the same equilibrium condition, Terzaghi's effective stress equation can be obtained, if the force induced by pore water pressure is excluded from the interparticle forces, as presented in Figure 5.2.

$$P = P_s' + u_w A \tag{5.4}$$

Then, the total stress and effective stress (excluding the interparticle stress induced by pore water pressure) can be represented by $\sigma_t = P/A$ and $\sigma = P_s'/A$, respectively. The effective stress equation can be therefore expressed as:

$$\sigma_t = \sigma + u_w \tag{5.5}$$

Actually, the same expression (Equation (5.5) can be directly derived by reorganizing the elements in Equation (5.2) to:

$$P = (P_s - u_w A_s) + u_w A = P_s' + u_w A \tag{5.6}$$

This illustrates that the effective stress, indeed, demonstrates the interparticle stress generated by the external forces other than pore water pressure. Based on such a definition of effective stress, the conclusion could not be drawn that Terzaghi's effective stress equation needs to be modified via interparticle force analysis.

Therefore, Terzaghi's effective stress equation would not need to be modified if effective stress was defined as the soil skeleton stress, excluding the effect of pore fluid pressure.

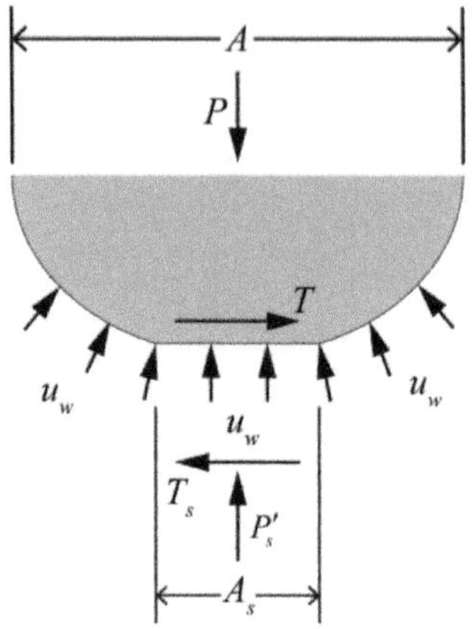

Figure 5.2 Pore water pressure and the force on the contact surface

However, if we must define that "the effective stress is the stress that governs the strength and deformation of soils," Terzaghi's effective stress equation will have to be modified when the effect of the pore fluid pressure on the strength and deformation of the soils is nonnegligible, which is also the reason for the modification. For saturated soils, the modified equation is Skempton's equation (Equations 3.8 and 3.9). Nonetheless, modifying the results in this way makes the effective stress a pseudo-physical quantity, which could not satisfy the equilibrium conditions for a free body of the soil skeleton or be directly substituted into the stress-strain constitutive equation. In fact, the definition of effective stress as the soil skeleton stress without the effect of pore fluid pressure does not affect us to consider the influence of pore fluid pressure on the strength and deformation of soils. When Terzaghi's effective stress is known, the influence of pore fluid pressure on shear strength and volumetric change can be taken into account with Equations (3.8) and (3.9). To put it simply, the strength and deformation of soils are actually the strength and deformation of the soil skeleton, respectively; both the effective stress and pore fluid pressure have contributions to the strength and deformation of the soil skeleton, while the effect of pore fluid pressure is usually ignorable.

Although the shear strength and volumetric deformation are entirely determined by the stresses deduced via the shear strength- and volumetric change-equivalency, these two stresses cannot be equivalent. Then which one is truly the effective stress? Or are they both effective stresses? Obviously, the latter statement is correct, following the definition of effective stress; i.e., both of them are effective stresses. This, however, raises another issue: which stress should we use in the equilibrium equation while solving the problems of soil consolidation and analysis of stress and deformation? It seems to be the stress derived from the volumetric change-equivalency. If this is the case, then is the stress derived from the shear strength-equivalency still effective stress?

Actually, it is inappropriate to use any of aforementioned stresses in the equilibrium equation, since neither of them satisfies the equilibrium condition of a free body of soil or the soil skeleton. Furthermore, it is less likely the two stresses satisfy the equilibrium condition simultaneously. Terzaghi's effective stress has to be the one in the equilibrium equation for saturated soils and the other porous media regardless of the circumstances and whether the contribution of pore water pressure to the strength and deformation needs to be considered. In other words, the equilibrium equation has to be Biot's consolidation equation.

Therefore, the logical order of causes and consequences actually would be inverted if effective stress was deduced with shear strength- or volumetric change-equivalency. Effective stress, as well as pore fluid pressure, is the cause, and shear strength and volumetric change are the consequences. The essential differences of the two causes would be confusing, leading to the complication and disorder of the studies on soil mechanics, if the "consequences" were used to deduce the "causes" while the "causes" were used to investigate the "consequences." Take the studies on soil constitutive relations, for example: it would be difficult to analyze and master the stress-strain law if the relationship between stress and strain was explored by using the stress deduced with volumetric change- or shear strength-equivalency. One of the most significant problems herein is that there are many impact factors for the strength and deformation of soils besides stresses, such as the structure of the soil skeleton and the physicochemical reaction between the mineral constituents of soil particles and water. The effect of these factors on the strength and deformation of soils, however, could not be simply reflected by stress variables.

However, the stress derived from volumetric change- and shear strength-equivalency indeed is a physical quantity without any other physical meaning except shear strength- or volumetric change-equivalency.

Certainly, this discussion does not mean that the stress deduced with volumetric change- and shear strength-equivalency is senseless. The equivalency method can be directly used to determine the shear strength or volumetric change of a soil when the total stress and pore fluid pressure are known, which omits the deduction processes and simplifies the problems.

It should be noted that the statement "effective stress controls the strength and deformation of soils" is different from "effective stress is the stress that governs the strength and deformation of soils." The former statement is more accurate, while the latter one leads to the impression that effective stress is equivalent stress. The former statement is conditional, where the influence of pore fluid pressure on the shear strength and volumetric change of soils could be negligible; while the result deduced with the latter one is an equivalent and pseudo-physical quantity.

Effective stress can be allowed to be define as equivalent stress in certain circumstances. In order to clarify the origins for deduction, it is better to define effective stress as the soil skeleton stress due to external forces and equivalent stress as the stress derived with volumetric change- and shear strength-equivalency. This clarification means deriving the effective stress equations with the original meaning of Terzaghi's effective stress in relatively correct approaches for soil mechanics studies.

How to define effective stress determines whether Terzaghi's effective stress equation needs to be modified. The two definitions reflect two different ways of thinking (i.e., phenomenological methods and rational methods) for solving soil mechanics problems. Shear strength and volumetric change are phenomena and consequences, for which effective stress is the cause. Deriving effective stress from volumetric change- and shear strength-equivalency can be generally classified as phenomenological methods, which would make the causes for the change in strength and volume confusing and the studies on soil mechanics complicated.

5.2 Is effective stress pseudo or real stress?

The "effective stresses" of soils obtained with shear strength- or volumetric change-equivalency, which are the stresses obtained with various modified Terzaghi's effective stress equations, are all pseudo stresses. Up to this point, there has been one perspective that considered "effective stress is a pseudo physical quantity and should be deduced with strength- or deformation-equivalency" in soil mechanics (Shen, 1995).

To answer the question of "Is effective stress pseudo or real stress?" the following perspectives have to be reaffirmed: (1) the strength and deformation of soils are just their counterparts of the soil skeleton; and (2) the strength and deformation are determined by the soil skeleton stress. Due to the second perspective, soil skeleton stress is naturally effective stress. However, as discussed in the early chapters the soil skeleton stress includes the part induced by the external forces and the other part induced by pore fluid pressure. Since the effect of the latter part on the strength and deformation of soils can normally be ignored, the soil skeleton stress due to external forces is called effective stress.

Soil skeleton stress is a measurement for the internal force of the soil skeleton (i.e., the internal force of the soil skeleton on a unit area). The internal force of the soil skeleton induced by external loading is real. Whether it should be averaged by the total area or skeleton area of soils is determined by the convenience for solving problems. Certainly, neither average reflects the true force in soil particles on a micro-scale. However, on a macro-scale, both averages are meaningful as measurements for the intensity of the internal force of the soil skeleton without any obstruction on the studies of the strength, deformation and stress-strain behavior of soils. It is easier to understand this by comparing to the definition of seepage velocity: the flux of soil seepage is real. Whether the definition of seepage velocity is the flux averaged by the total area or the pore area of soils is determined by the convenience of the studies. Similarly, regardless of the fact that neither definition reflects the real velocity of pore water flow, the seepage model we set up with the flow velocity over section area can still solve the seepage problems of soils accurately.

In fact, it can be found that the definition of seepage velocity is entirely consistent to that of stress with the aid of continuous medium models of soils. This can further elucidate the significance of establishing a continuous medium model on soil mechanics studies.

Using the stress over the total area to set up equilibrium differential equations is more convenient, while it is easier to use the stress over skeleton area for the stress-strain constitutive relations of soils.

We can definitely insist the average stress is a virtual quantity, just as the seepage velocity does not represent the real velocity of pore flow. However, this virtualization is substantially different from the pseudo stress of soil derived from shear strength- or volumetric change-equivalency. The later stress can neither satisfy equilibrium equations nor describe the stress-strain properties of soils essentially. On the other hand, as an artificial definition, the effective stress can be defined as soil skeleton stress due to external forces, or strength- or volumetric change-equivalent stress. Nevertheless, soil skeleton stress objectively exists and governs the deformation and strength of the soil, whether it shares the concept of effective stress or not, or, say, whether the effective stress is defined as the soil skeleton stress due to external forces or not.

5.3 Effective stress and stress state variables of soils

The macro physical quantity describing the stress state of a point in soils is called a stress state variable, which is a stress variable that controls the structure equilibrium of soil structure. In unsaturated soil mechanics, the stress state variable is specified as the independent stress element that controls the strength and deformation of unsaturated soils.

Fredlund and Morgenstern (1977) were the first to clearly introduce the concept of stress state variable, in order to overcome the difficulties in seeking a reasonable expression of effective stress to control strength and deformation for unsaturated soils. They obtained the corresponding equilibrium differential equations via a force analysis on free bodies of soil mass, pore water, pore air and contractile skin. Based on the equilibrium differential equations, they pointed out the stress state variables that control the deformation and strength of unsaturated soils are the combination of any two stresses of $(\sigma_t - u_a)$, $(\sigma_t - u_w)$ and $(u_a - u_w)$. This thought was widely accepted in the soil mechanics community and became an important research direction for unsaturated soils, even leading most people in this field to abandon the concept of effective stress for unsaturated soils. Then, for unsaturated soils, is there effective stress that controls their strength and deformation? The answer is "Yes" based on our previous discussion in this book. We could not imagine that effective stress does exist for saturated soils but not for unsaturated soils, with lower water content.

What is the relationship between the effective stress and stress state variables in unsaturated soils? Effective stress itself is a stress state variable, and also a combination of the specific stress state variables. Based on its expression, effective stress is the combination of net stress u_w $(\sigma_t - u_a)$ and matric suction $(u_a - u_w)$. It can also be the combination of the other stress state variables when the expression is modified, such as $(\sigma_t - u_w)$ and $(u_a - u_w)$. The coefficient in the combination is the effective degree of saturation. Actually, based on the principle that the stress state variables have to be provided by the equilibrium differential equation of internal forces (Fredlund and Morgenstern, 1977), stress state variables can also be derived from the equilibrium equation expressed by effective stress, of which the forms are $\{\sigma, u_a, u_w\}$, $\{\sigma, u_a, u_a - u_w\}$ or $\{\sigma, u_a, Se\,(u_a - u_w)\}$, where σ is the effective stress.

The concept of a stress state variable is broadly applicable in solving shear strength problems and establishing constitutive models, while effective stress is the specific stress variable that controls the strength and deformation of soil. This means that the strength and deformation characteristics of soil can always be described by an appropriate stress state variable, regardless of the soil's type and surrounding conditions. That said, the characteristics of strength and deformation of a soil can always be expressed as a combination of several stress state variables, whereas the effective stress is a specific combination of a series of stress state variables. Once the effective stress is obtained, it is unnecessary to seek the other combinations of stress state variables for describing the strength and deformation of the soil. In other words, the studies on stress state variables do not directly tell us which combination can control the strength and deformation of the soil. In other words, how to combine the variables describing the strength and deformation properties of soils is still unknown even though all the stress state variables are available. Therefore, the correct combination forms have to be explored with experiments. However, effective stress explicitly provides the specific combination of stress state variables that governs the strength and deformation of unsaturated soils.

For unsaturated soil mechanics, the studies on stress state variables are no longer significant or urgent once the effective stress is determined. As mentioned earlier, the studies on stress state variables only specify the combination term of stresses rather than the specific form of the combination that controls the strength and deformation of soils. On the contrary, the effective stress determines both the stress state variables and their specific combinations. Specifically, the strength and volumetric change of soils can be exclusively determined by the effective stress when the influence of pore fluid pressure is negligible. The formulae are:

$$\tau_f = c' + \{\sigma_t - [u_a - S_e(u_a - u_w)]\}\tan\varphi' \tag{5.7}$$

and:

$$-\frac{\Delta V}{V} = C\cdot\{\Delta\sigma_t - \Delta[u_a - S_e(u_a - u_w)]\} \tag{5.8}$$

Furthermore, if the contribution of pore fluid pressure to the strength and volumetric change of soils has to be considered, the corresponding formulae are:

$$\tau_f = c' + \left\{\sigma_t - (1 - \frac{a_c \tan\psi}{\tan\varphi'})[u_a - S_e(u_a - u_w)]\right\}\tan\varphi' \tag{5.9}$$

and:

$$-\frac{\Delta V}{V} = C\cdot\left\{\Delta\sigma_t - \left(1 - \frac{C_s}{C}\right)\Delta[u_a - S_e(u_a - u_w)]\right\} \tag{5.10}$$

In the end, it has to be illustrated about the stress state variables: due to the multiple phase nature of a soil, the stress state variables are related to the study objects based on their definition. The specific study objects and their corresponding stress state variables are listed in Table 5.1.

It should be noted that the stress variables for the soil skeleton in unsaturated soils has other forms besides those shown in Table 5.1, which are the combination of $(\sigma_t - u_a)$ and $(u_a - u_w)$; σ_t, u_a and u_w; and etc.

5.4 Effective stress and soil skeleton stress

The effective stress is determined by the equilibrium equation of the soil skeleton; so are the stress state variables (Fredlund and Rahardjo, 1993). However, why did the previous researchers never obtain the effective stress in this way? The key is the definition of the soil skeleton stress. In introducing the stress state variables, Fredlund did not directly present the equilibrium differential equation for the soil skeleton, nor did he discuss the definition of soil skeleton stress. He treated soil as a four-phase object consisting of the soil skeleton,

Table 5.1 The study objects and their corresponding stress state variables

Study objects	*Stress state variables*
Soil mass	σ_t
Soil skeleton (unsaturated soils)	$(\sigma_t - u_a)$ and $(u_a - u_w)$
Soil skeleton (saturated soils)	σ_t and u_w
Pore water	u_w
Pore air	u_a

contractile skin, pore water and pore air. The equilibrium differential equation for the soil skeleton was derived by subtracting those for pore water, pore air and contractile skin from the equilibrium equation for total stress rather than the force analysis and equilibrium condition of the soil skeleton, as illustrated in Section 5.6. More researchers determined the equilibrium differential equation of the soil skeleton following the theory of mixtures or the definition of skeleton stress in porous medium mechanics.

Skeleton stress is usually defined with the Cauchy stress tensor in the theory of mixtures and porous medium mechanics (Olivier, 2004; Reint, 2005). Specifically, the skeleton stress is defined as the internal force over skeleton area and introduced by all the external forces (volume forces and surface forces) at one point of soil in one direction. The pore fluid pressure is included in these external forces.

Carroll (1980) proposed the average stress theorem, i.e., the definition of stress is:

$$\overline{\sigma_{ij}} = \frac{1}{V}\int_V \sigma_{ij} dV = \frac{1}{V}\left[\int_{\partial V} t_i X_i dA + \int_V \rho b_j X_i dV\right] \tag{5.11}$$

where V is the volume of the soil body, ∂V is the boundary of V, $\overline{\sigma_{ij}}$ is the average stress, σ_{ij} is the stress at one point, t_i is the surface force component, b_j the volume force component and X_i the coordinate value.

The average stress theorem elucidates that the average stress of the soil body is determined by the loading and geometrical properties of the deformable body, and is established for either linear or nonlinear constitutive relations depending on whether the materials are homogeneous, but is unrelated to the materials' response.

However, if the effective stress was defined as the skeleton stress due to external stresses, i.e., the component due to pore fluid pressures was excluded in the definition, the effective stress-expressed equilibrium equation of the soil skeleton could actually be deduced with the method of porous medium mechanics. The effective stress equation can further be obtained. The deduction is indicated as follows:

A section of a representative volumetric element of the soil skeleton (RVE) is shown in Figure 5.3. The total internal force of the section is designated as $\Delta\vec{F}$, and the vertical and horizontal components are $\vec{N}$ and $\vec{T}$, with the values of N and T, respectively. Therefore, the

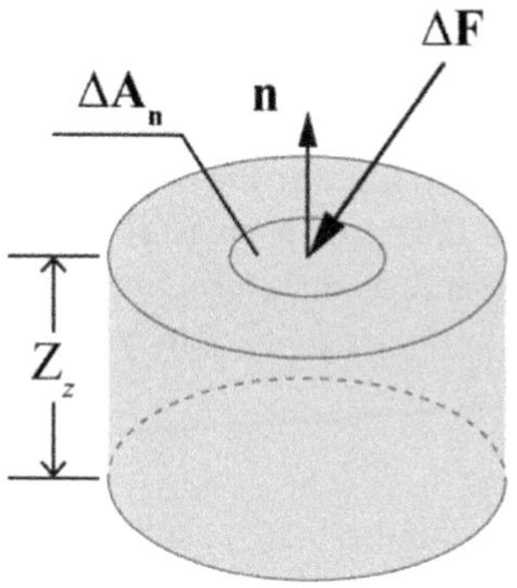

Figure 5.3 The internal force on the total area and the soil skeleton area

soil skeleton stress in the continuous medium mechanics and porous medium mechanics is defined as:

$$\sigma^{st} = \frac{N}{A}, \tau^{st} = \frac{T}{A} \tag{5.12}$$

or:

$$\sigma^{s} = \frac{N}{(1-n)A}, \tau^{s} = \frac{T}{(1-n)A} \tag{5.13}$$

where A is the total area of the section, $(1 - n)A$ is the area of the soil skeleton and n is the porosity. Equation (5.12) indicates the internal force of the soil skeleton over the total area, while Equation (5.13) represents that over the soil skeleton area.

Note that pore water pressure and pore air pressure have contributions on N but not on T. Herein, we eliminate the internal force $\vec{N}_w$ and $\vec{N}_a$, respectively, from $\vec{N}$ and define the normal **effective** stress as:

$$\sigma = \frac{N - N_w - N_a}{A} \tag{5.14}$$

where $\vec{N}_w$ and $\vec{N}_a$ are the skeleton internal forces induced by pore water pressure and pore air pressure, respectively, and their corresponding values are N_w and $\vec{N}_a$.

According to the force analysis of the free body of the soil skeleton in Chapter 2, we know:

$$N_w = u_w(1-n)\frac{n_w}{n}A, N_a = u_a(1-n)\frac{n_a}{n}A \tag{5.15}$$

where n_w and n_a are the pore water pressure and pore air pressure, of which the corresponding porosities are n_w and n_a, respectively, and $n_w + n_a = n$.

The relationship between the soil skeleton stress and effective stress can be obtained based on the definition of soil skeleton stress in Equations (5.12) and (5.13), and the definition of effective stress in Equation (5.14). This relationship can also be obtained according to the force analysis on the surface of the free body of the soil skeleton. Take the definition of the soil skeleton stress of Equation (5.13) as an example.

The soil skeleton stress σ^s and the applied area are schematically shown in Figure 5.4(a), where N represents the normal internal force on the section of the soil skeleton; in Figure 5.4(b), the σ, the pore water pressure and the pore air pressure and their applied areas are presented, respectively, where N means the same as that in Figure 5.4(a).

The internal force N shown in Figure 5.4(a) can be expressed as:

$$N = \sigma^s(1-n)A \tag{5.16}$$

The internal force N shown in Figure 5.4(b) can be expressed as:

$$N = \sigma A + (1-n)\frac{n_w}{n}Au_w + (1-n)\frac{n_a}{n}Au_a \tag{5.17}$$

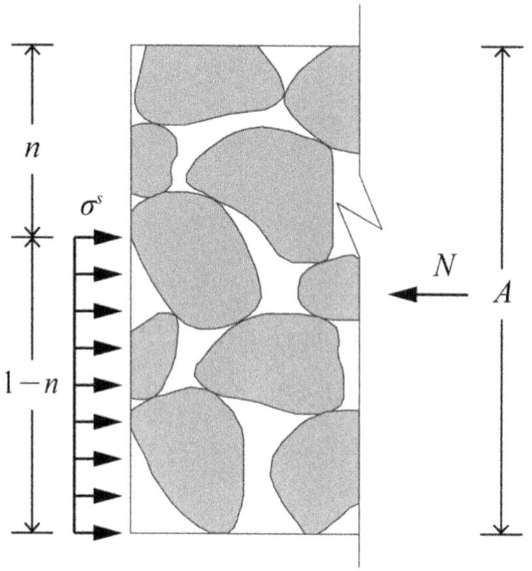

(a) The soil skeleton stress σ^s and its action area

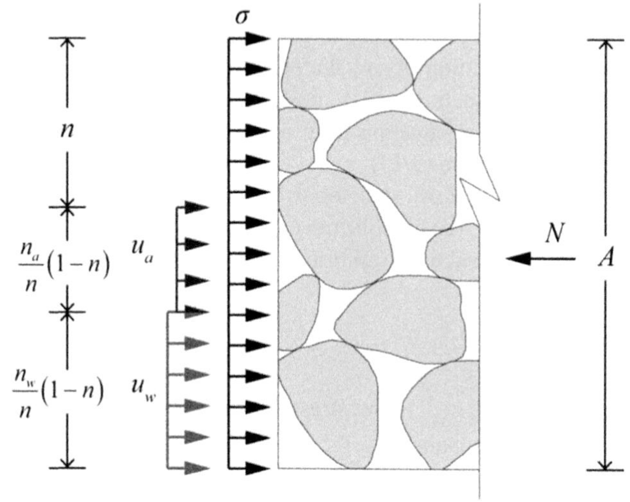

(b) The effective stress, pore water pressure, pore air pressure and their action areas

Figure 5.4 The average stress theorem-defined soil skeleton stress and effective stress, and their action areas

Since N in Equation (5.16) is equal to that in Equation (5.17), it can be derived from these two equations that:

$$(1-n)\sigma^s = \sigma + (1-n)\frac{n_w}{n}u_w + (1-n)\frac{n_a}{n}u_a \tag{5.18}$$

It should be noted that:

$$\frac{n_w}{n} = S, \frac{n_a}{n} = 1-S \tag{5.19}$$

where S is the degree of saturation, and thus Equation (5.18) can be written as:

$$(1-n)\sigma^s = \sigma + (1-n)Su_w + (1-n)(1-S)u_a \tag{5.20}$$

or:

$$\sigma = (1-n)\sigma^s - (1-n)Su_w - (1-n)(1-S)u_a \tag{5.21}$$

The soil skeleton stress equation (5.15), which is defined with the average stress theorem, is different from the effective stress in two aspects: first, it is the average of all the external forces on the soil skeleton rather than that excluding pore fluid pressure; second, it is averaged by the soil skeleton area instead of the area of soil mass.

The difference in the definition of soil skeleton stress determines the difference among the equilibrium differential equations. It is illustrated in the following how to derive the expression of the effective stress equation from equilibrium differential equations based on the soil skeleton stress in Equation (5.13).

According to the definition of soil skeleton stress in Equation (5.13), without considering the effect of soil skeleton water, the equilibrium equation of the soil skeleton can be obtained via the internal force analysis and the equilibrium condition of the free body of the soil skeleton (Chen and Qin, 2012):

$$(1-n)\sigma^s_{ij,j} + (1-n)\rho_s g_i + f_i^{sw} + f_i^{sa} = 0 \tag{5.22}$$

where n is the porosity of soil, $\sigma^s_{ij,j}$ is the stress of the soil skeleton, $(1-n)\rho_s g_i$ is the soil skeleton gravity on unit volume, and f_i^{sw}, f_i^{sa} are the seepage force of water and air on the soil skeleton.

The equilibrium equations for pore water and pore air are:

$$nSu_{w,i} + nS\rho_w g_i - f_i^{ws} = 0 \tag{5.23}$$

$$n(1-S)u_{a,i} + n(1-S)\rho_a g_i - f_i^{as} = 0 \tag{5.24}$$

where S is the degree of saturation, $u_{w,i}$ and $u_{a,i}$ are the pore water pressure and pore air pressure, $nS\rho_w g_i$ and $n(1-S)\rho_a g_i$ represent the gravity of water and air on the unit volume of soil, respectively, and f_i^{ws} and f_i^{as} represent the force on the water and air applied by the soil skeleton.

Substituting the relationship between σ^s and σ in Equations (5.20) and (5.21) into Equation (5.22) can obtain:

$$\sigma_{ij,j} + (1-n)Su_{w,i} + (1-n)(1-S)u_{a,i} + (1-n)\rho_s g_i + f_i^{sw} + f_i^{sa} = 0 \tag{5.25}$$

Adding the equilibrium equations of the soil skeleton, pore water and pore air, i.e., Equations (5.25), (5.23) and (5.24), with the mass of pore air neglected, we have:

$$\sigma_{ij,j} + Su_{w,i} + (1-S)u_{a,i} + \rho g_i = 0 \tag{5.26}$$

Comparing Equation (5.26) to the total stress equation for unsaturated soils (Equation (2.2)) can derive the relationship among the total stress, effective stress and pore fluid pressure, i.e., the effective stress equation for unsaturated soils:

$$\sigma_{ij} = \sigma_{tij} - Su_w\delta_{ij} - (1-S)u_a\delta_{ij} \tag{5.27}$$

which can be further developed into:

$$\sigma_{ij} = \sigma_{tij} - u_a\delta_{ij} + S(u_a - u_w)\delta_{ij} \tag{5.28}$$

It can be found that Equation (5.28) is the effective stress equation for unsaturated soil. According to the deduction procedure of Equation (5.28), it can be known that the equilibrium differential equation for the soil skeleton and the effective stress equation for unsaturated soils can be derived from the equilibrium condition of a free body of the soil skeleton, as long as the definition of effective stress is accepted. In addition, Equation (5.28) can always be obtained no matter whether the soil skeleton stress is defined as the internal force on the unit area of soil mass or that of the soil skeleton.

In fact, the expression of effective stress can also be obtained through the internal force analysis on a free body of soil mass, which can be demonstrated with the normal stress in the horizontal direction, as schematically shown in Figure 5.5.

The total internal force in the horizontal direction on the vertical section is designated by N_t, and then:

$$\sigma_t = \frac{N_t}{A} \tag{5.29}$$

where A is the total area of the soil section.

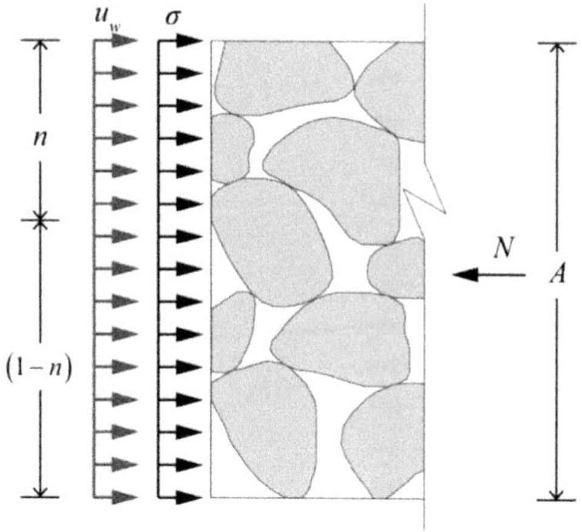

Figure 5.5 The force analysis of a free body of soil mass in the horizontal direction

Assuming the effective normal stress in the horizontal direction is σ and the pore water pressure is u_w, then the internal force of the soil skeleton due to pore water pressure is $u_w(1-n)A$ and the fluid pressure on the pore water is $u_w nA$. The relationship of the internal forces in the horizontal direction is:

$$N_t = \sigma A + u_w(1-n)A + nu_w A \tag{5.30}$$

and thus:

$$\sigma_t = \sigma + u_w \tag{5.31}$$

which is the Terzaghi's effective stress equation for saturated soils.

Similarly, the horizontal forces on a free body of unsaturated soil are shown in Figure 5.6. Here, σ_x and τ_{xz} represent the normal stress and shear stress on the soil skeleton due to the external forces, excluding pore fluid pressure, i.e., the soil skeleton stress due to external forces. The definition of this stress is the internal force on unit soil area resulting from all the external forces, excluding pore fluid pressure. Simultaneously, there is also the skeleton stress due to pore water pressure and pore air pressure on the section of the soil skeleton. Letting u_w and u_a denote the values of the stresses on the soil skeleton due to pore water pressure and pore air pressure, and A_{sw} and A_{sa} as applied areas of pore water pressure and pore air pressure, respectively, we have:

$$A_{sw} = \frac{n_w}{n} A_s \tag{5.32}$$

$$A_{sa} = \frac{n_a}{n} A_s \tag{5.33}$$

where $A_s = (1-n)A$, $A_{sw} + A_{sa} = A_s$, A_s is the total section area of soil mass, A_s is the area of the soil skeleton, n is the porosity, n_w and n_a are the porosities corresponding to pore water

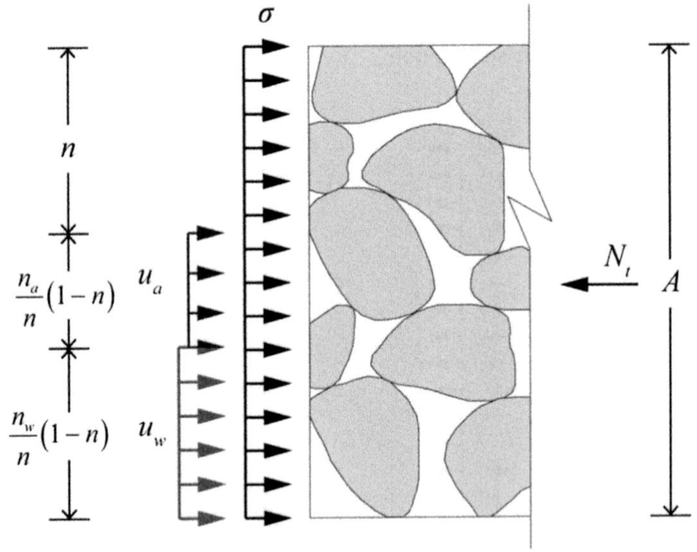

Figure 5.6 The force analysis of a free body of soil mass in the horizontal direction for unsaturated soils

and pore air, respectively, and n_w/n and n_a/n are the fractions of pore water and pore air on the area occupied by pore fluid, respectively.

The total normal internal forces of soil mass and the soil skeleton are expressed by N_t and N_s, respectively:

$$N_t = N_s + N_w + N_a \tag{5.34}$$

and:

$$N_w = u_w nA \frac{n_w}{n} = u_w n_w A \tag{5.35}$$

$$N_a = u_a nA \frac{n_a}{n} = u_a n_a A \tag{5.36}$$

where nA is the pore area and N_w and N_a designate the internal forces due to pore water pressure and pore air pressure.

According to the analysis on the external force-induced skeleton stress and pore pressure-induced skeleton stress, the normal internal force on the section of the soil skeleton is found to be the sum of the internal forces corresponding to the aforementioned stresses, i.e.:

$$N_s = \sigma A + u_w (1-n) \frac{n_w}{n} A + u_a (1-n) \frac{n_a}{n} A \tag{5.37}$$

The expression of N_t is:

$$N_t = \sigma_t A \tag{5.38}$$

where σ_t is the total stress. Substituting Equation (5.35), (5.36), (5.37) and (5.38) to (5.34):

$$\sigma_t = \sigma + u_w \frac{n_w}{n} + u_a \frac{n_a}{n} \tag{5.39}$$

It should be noted that $\frac{n_w}{n} = S$ and $\frac{n_a}{n} = 1 - S$, therefore:

$$\sigma_t = \sigma + S u_w + u_a (1 - S) \tag{5.40}$$

i.e.:

$$\sigma = \sigma_t - u_a + S(u_a - u_w) \tag{5.41}$$

When the pore water corresponding to residual water content is considered as a part of the soil skeleton, Equation (5.41) should be revised to:

$$\sigma = \sigma_t - u_a + S_e(u_a - u_w) \tag{5.42}$$

where S_e is the effective degree of saturation.

Equation (5.42) is the effective stress equation for unsaturated soils, which is directly obtained based on the relationship among the internal forces on the section of soil mass.

It has to be confirmed that the soil skeleton stress based on the average stress theorem in porous medium mechanics is a real measurement with an explicit physical meaning. If not for the different effects resulting from pore fluid pressure and the other external forces on the soil skeleton, these two types of effects would not have to be distinguished, and thus the skeleton stress given by the average stress theorem can be directly applied to describe the strength and stress-strain properties of the soil skeleton. However, this difference is all known and has been repeatedly declared in previous texts. Therefore, it is more convenient to differentiate these two types of effects. On the other hand, is it meaningful to separately define the soil skeleton stresses by distinguishing these two types of effects? The answer is that it is not only significant, but also much easier for indicating the physical meaning of Terzaghi's effective stress thanks to this definition, and more straightforward for understanding the physical meaning of shear strength- or volumetric change-equivalent "effective stress."

Meanwhile, for the definition of soil skeleton stress, is it more appropriate to average the skeleton internal force by the soil skeleton area or the soil mass area? This question is the same as that whether the seepage velocity is the flux averaged by pore area or soil mass area. The answer is that it is determined by which way is more convenient for solving the problems. From the authors' perspective, it is more convenient to define the soil skeleton stress as the internal force of the soil skeleton over the soil mass area for establishing equilibrium equations for stress analysis; whereas the definition of the internal force of the soil skeleton over the soil skeleton area is more feasible for studying the stress-strain constitutive relationship of soils. In addition, stresses are all defined with soil mass area in traditional soil mechanics. Both the total stress and effective stress are defined as the average of internal force on soil mass area. Therefore, it is better to define the soil skeleton stress as the internal force of the soil skeleton on the unit soil mass area to be consistent with traditional soil mechanics. Furthermore, the unit soil mass area is the total area of the space occupied by the soil skeleton. Therefore, using the internal force of the soil skeleton on the unit soil mass area as the definition of the soil skeleton stress is also consistent to the definition of stress in continuous medium mechanics.

5.5 The effective stress of unsaturated soils

To keep using the traditional concept of effective stress, we defined it as the soil skeleton stress due to the external forces, excluding pore fluid pressure. Based on this definition, an invariant relationship is established among effective stress, total stress and pore fluid pressure, which is the effective stress equation. Meanwhile, we also modified the perspective that the effective stress should be derived based on the shear strength- and volumetric change-equivalency of soils. We revised it to: (1) the shear strength and volumetric change of soils is related to effective stress and pore fluid pressure; (2) usually the effective stress governs the shear strength and deformation of soils, on which the effect of pore fluid pressure can be ignored.

The effective stress concept and effective stress equation are applicable for all types of soils, no matter whether they are dry soils or moisturized soils, saturated soils or unsaturated soils. It should be noted that there is only pore air pressure when the soil is dry. Meanwhile, the effective stress equation is irrelevant to the properties of the interparticle forces of soils (e.g., long-range force, short-range force, electric suction and van der Waals force). The analysis found that the deduction of the equilibrium equation does not need the pore fluid to be interconnected. When it is unconnected, the pore fluid may also result in pressure on the soil skeleton. The effective stress equation is established as long as the pore fluid pressure exists. Therefore, the effective stress equation is applicable for soil with various water content, although the physical and mechanical properties of dry soil and moisturized soil (i.e., the soils with the water content lower and higher than the residual water content, respectively) are significantly different. There is no need to distinguish the soils to water interconnected soil, air interconnected soils or water-air connected soil. From the opposite standpoint, if the effective stress equation is only applicable for the soils with the water content higher than a specific value, this value is the threshold water content, which must have an explicit physical meaning. The residual water content has all these characteristics. The clayey soils would turn from unsaturated soils to dry soils once their water content is lower than the residual water content. With the decrease of water content, the bonding water film would become thinner, and thus the structure and properties of soil mass would change.

The pores in the natural unsaturated soil layers are generally connected to the atmosphere. Therefore, the pore air pressure is equal to the relative atmospheric pressure, of which the value is fairly close to 0. The value of matric suction is related to the degree of saturation. The matric suction is relatively small when the degree of saturation is relatively large; by contrast, the matric suction is relatively large when the degree of saturation is relatively small. Since the degree of saturation (water content) determines the action area of pore water pressure (matric suction), the influence of pore fluid pressure on the strength and deformation can be neglected, which can be totally determined by effective stress in general cases.

Saturated and unsaturated soils have a uniform effective stress equation. The equation for saturated soils is the special case of that for unsaturated soils. Therefore, the strength and deformation problems of saturated soils and those of unsaturated soils can be unified. Correspondingly, the problems of stress, deformation and stabilization analysis of soil structures for saturated soils and those for unsaturated soils can be unified.

For unsaturated soils, the form of the effective stress equation we derived is the same as Bishop's effective stress equation, although it is irrelevant to the definition of effective stress and the deduction of the effective stress equation. Note that the coefficient in Bishop's formula should be the effective degree of saturation. The effective stress equation for unsaturated soils should be obtained through the comparison between the equilibrium equation expressed by the effective stress of soils and that expressed by total stress. In this book, the effective stress is defined as the soil skeleton stress induced by the external forces, excluding pore fluid pressure. This type of definition is the inevitable result in the force analysis of a free body of the soil skeleton. As aforementioned, the effect of pore fluid pressure on the soil skeleton would be revealed when we used the soil skeleton as the study subject to conduct force analysis on the free body of the soil skeleton. The effect of uniform pore fluid pressure on the soil skeleton is different from that of the other external forces. These two sorts of effects can be considered separately. In this circumstance, the effect of the other external forces is just Terzaghi's effective stress for saturated soils. Therefore, it is undoubtedly a rational choice to use the concept of effective stress, or in other words, to use the stress induced by the external forces except pore fluid pressure to interpret effective stress.

For saturated soils, the feasibility of the effective stress equation has been sufficiently confirmed. However, for unsaturated soils, whether the effective stress is applicable regardless of water content still needs further verification. Experimental data showed that there is no definite linear relationship between the coefficient in Bishop's formula and the degree of saturation. The authors believe that the comparative complication of the experiments for unsaturated soils is responsible to this problem. In the experiment processes, soil porosity might vary. Correspondingly, the water content, residual water content and thus the soil-water characteristic could vary. All the variation in experimental measurement, data organization and analysis are difficult to consider accurately. Moreover, the accuracy of the measurement for the deformation of soil samples is usually difficult to satisfy the experimental requirement. Therefore, the current limited experiment data is insufficient to dismiss the applicability of effective stress equation. On the other hand, after the proposal of the dual-stress variable controlled shear strength equation for unsaturated soils, Vanapalli and Fredlund (Vanapalli *et al.*, 1996) obtained the shear strength expression for unsaturated soils via experiments and verified that the controlling factor for the shear strength of soils is the effective stress. Due to the hint on the shear strength experiment results of unsaturated soils given by Vanapalli and Fredlund, we considered the pore water corresponding to residual water content as a constituent of the soil skeleton, based on the investigation of the definition of the soil skeleton. Lu *et al.* (2010) proposed the concept of suction stress and further derived the effective stress equation with the same form as that in this book, and confirmed the effectiveness of the effective stress equation by referring a great amount of shear strength experiment data of unsaturated soils.

The meaning of declaring the effective stress of unsaturated soils is that we can unify the mechanical problems of saturated soils and unsaturated soils and establish a knowledge system for unsaturated soil mechanics following the counterpart for saturated soils, after deriving the equilibrium differential equations and effective stress equations for unsaturated soils.

5.6 Should contractile skin be the fourth phase?

Unlike for saturated soils, the effective stress that controls the strength and deformation of unsaturated soils has not been proposed, and hence many researchers gave up the effort in exploring the effective stress of unsaturated soils. They, in turn, sought an appropriate stress state variable to characterize the stress state of unsaturated soils and studied the strength and deformation. Fredlund has made great contributions in the investigation of unsaturated soil mechanics. He is also among the earliest researchers who studied the stress state variable of unsaturated soils.

Fredlund determined the stress state variable based on the equilibrium equation for unsaturated soils. However, in the derivation of the equilibrium equation for unsaturated soils, he did not directly take the free body of the soil skeleton to conduct the force analysis. Instead, Fredlund used pore water and pore air as the analysis subjects and considered the surface contractile skin of unsaturated soils as an independent fourth phase. After obtaining the equilibrium differential equations for pore water, pore air and contractile skin, he derived the equilibrium differential equation for the soil skeleton by substituting the counterparts of the previous three phases from the equilibrium equation for the total stress of soil mass. Then Fredlund and Morgenstern (1977) determined the stress state variables of unsaturated soils based on the equilibrium equation for the soil skeleton.

In contrast to Fredlund's method, we did not consider the contractile skin as an independent phase. We conducted the internal force analysis on the free body of the soil skeleton directly and derived the equilibrium differential equation for the soil skeleton. Therefore, is it reasonable and necessary to treat the contractile skin as the fourth phase in force equilibrium analysis?

In physical chemistry, a medium phase is an aggregating state of the homogeneous matters with the same physical properties, chemical properties and composites without any external force. The gas in the system is always one single phase no matter if it is pure gas or mixed gas. If there is only one type of fluid, it is always one single phase, no matter whether this fluid is pure medium or (real) solution.

The knowledge basis for Fredlund and Morgenstern (1977) to consider the contractile skin as an independent phase in the force equilibrium analysis on the free body is the definition of phases and the study achievement on gas-water interface, i.e., contractile skin, by previous researchers. The main achievement is that "most properties of contractile skin are different from those of the adjacent water phase (also called essential body), e.g., smaller density, larger thermal conductivity (factor), and ice-analogical birefraction data. The transformation procedure from liquid water to contractile skin is apparent and mutational." It seems the study results could sufficiently indicate that the contractile skin is an independent phase state. Nonetheless, we have to know that (1) the properties of any material in the transition interface are different from that of the essential body; (2) the structure and motion feature (existing feature) of water molecules in contractile skin are not substantially different from those of the essential body; (3) the properties variation of the water in contractile skin is due to force change. In other words, the force on the water molecules in contractile skin and that in the essential body are different. The attractive force on the water molecules in the contractile skin applied by the gas molecules at air-water interface is smaller than the attractive force among the water molecules in the essential body; and (4) the contractile skin is formed due to force change instead of phase change. The motion of water molecules includes thermal motion (Brownian motion) and intermolecular attraction. The former motion results in water pressure, and the later induces cohesion. The cohesion forms contractile skin at interfaces and surface tension at the cross interfaces among skeleton, water and air. The amount of surface tension can be calculated with the difference between the pressures inside and outside contractile skin. Therefore, the contractile skin is the reflection of the molecular thermal motion and attraction at the solid-liquid-gas interfaces and the results of the soil skeleton–pore water–pore air interactions. From this point of view, it is more rational to consider the contractile skin as surface or interface, while the physics basis is not adequate for considering the contractile skin as an independent phase.

In fact, the contractile skin is only as thick as certain layers of water molecules. It would also be difficult to determine the interface among the contractile skin, water and air if it was treated as an independent phase.

Certainly, some researchers considered the interfaces between different phases as independent phases on which to conduct force equilibrium analysis. If necessary, it is not inappropriate to treat the contractile skin as an independent phase in force equilibrium analysis. We can even take any part of soil to conduct force equilibrium analysis independently. The question is whether it is necessary.

The interface or surface of homogeneous materials containing the same composites usually possesses different properties compared to the internal materials (essential body), e.g., the surfaces of solid metals. However, we do not have to conduct force equilibrium analysis on them independently. The objective of Fredlund is clear by treating the contractile skin as an independent phase in the force equilibrium analysis. He aimed to explore and indicate the stress state variables for unsaturated soils. Actually, doing so can also elucidate that the combination of $(\sigma_t - u_a)$ and $(u_a - u_w)$, and the combination of $(\sigma_t - u_w)$ and $(u_a - u_w)$ are the stress state variables for unsaturated soils by conducting internal force analysis on the soil skeleton phase, pore water phase and pore air phase separately. In this way, it is simpler and can directly obtain the expression of effective stress for unsaturated soils. In this case, there is no need to treat the contractile skin as an independent fourth phase.

It is illustrated in the following that the equilibrium equation of the soil skeleton derived via Fredlund's investigation on the four phases of the soil skeleton, contractile skin, pore water and pore air is consistent to the soil skeleton stress equation obtained through the force equilibrium analysis on the three phases of the soil skeleton, pore water and pore air in this book. Fredlund's equilibrium differential equation herein is referred to the article "Stress state variables for unsaturated soil," where the direction of the coordination was altered for comparison. In this paper, the equilibrium of a cubical element is considered within the context of continuum mechanics applied to a multiphase system. An element of unsaturated soil can be disassembled into a water, air, soil particle, and contractile skin phase. A linear equilibrium equation can be written for each of the four phases. In the vertical direction, an overall or total equilibrium equation can also be written as:

$$\frac{\partial \tau_{xz}}{\partial x}+\frac{\partial \tau_{yz}}{\partial y}+\frac{\partial \sigma_{tz}}{\partial z}+\rho g=0 \tag{5.43}$$

in which ρ = total density of saturated soil, σ_{tz} = total stress in z direction; τ_{xz} = shear stress on x-plane in z direction; τ_{yz} = shear stress on y-plane in z direction; and dx, dy, dz = unit dimensions of the element.

The equilibrium for the water phase in the z direction is:

$$n_w\frac{\partial u_w}{\partial z}+n_w\rho_w g+F_{sz}^w+F_{cz}^w=0 \tag{5.44}$$

in which u_w = pressure in the water phase; n_w = porosity with respect to the water phase; F_{sz}^w is the interaction force between the water and the soil particles; F_{cz}^w is the interaction force between the water and the contractile skin; and $n_w\rho_w g$ = gravity body force for the water phase.

The equilibrium for the air phase in the z direction is:

$$n_a\frac{\partial u_a}{\partial z}+n_a\rho_a g+F_{sz}^a+F_{cz}^a=0 \tag{5.45}$$

in which u_a = pressure in the air phase; n_a = porosity with respect to the air phase (i.e., percentage of the surface of the element going through air); $n_a\rho_a g$ = gravity body force for air phase; F_{sz}^a is the interaction force between the air and the soil particles; and F_{cz}^a is the interaction force between the air and the contractile skin.

The equilibrium equation for contractile skin is:

$$-n_c\frac{\partial f^*}{\partial z}(u_a-u_w)-n_c f^*\frac{\partial (u_a-u_w)}{\partial z}+n_c\rho_c g-F_{cz}^w-F_{cz}^a=0 \tag{5.46}$$

The equilibrium equation for soil skeleton structure can be obtained by subtracting the equilibrium equations for water phase, air phase and contractile skin phase from the total equilibrium equation, as:

$$\begin{aligned}&\frac{\partial \tau_{xz}}{\partial x}+\frac{\partial \sigma_{tz}}{\partial z}-n_a\frac{\partial u_a}{\partial z}-n_w\frac{\partial u_w}{\partial z}+n_c f^*\frac{\partial (u_a-u_w)}{\partial z}+\frac{\partial \tau_{yz}}{\partial y}+\rho g\\&-n_a\rho_a g-n_w\rho_w g-n_c\rho_c g-F_{sz}^w-F_{sz}^a+n_c(u_a-u_w)\frac{\partial f^*}{\partial z}=0\end{aligned} \tag{5.47}$$

The equilibrium equation for the soil skeleton with contractile skin can be obtained by adding Equations (5.46) and (5.47), i.e., the equilibrium equations for contractile skin and the soil skeleton, as:

$$\frac{\partial \tau_{xz}}{\partial x}+\frac{\partial \sigma_{tz}}{\partial z}-n_a\frac{\partial u_a}{\partial z}-n_w\frac{\partial u_w}{\partial z}+\frac{\partial \tau_{yz}}{\partial y}+\rho g-n_a\rho_a g-n_w\rho_w g-n_c\rho_c g$$
$$-F_{sz}^{w}-F_{sz}^{a}-F_{cz}^{w}-F_{cz}^{a}=0 \tag{5.48}$$

It can be reorganized to

$$\frac{\partial \tau_{xz}}{\partial x}+\frac{\partial \tau_{yz}}{\partial y}+\frac{\partial \sigma_{tz}}{\partial z}-n_a\frac{\partial u_a}{\partial z}-n_w\frac{\partial u_w}{\partial z}-\left(F_{sz}^{w}+F_{cz}^{w}\right)-\left(F_{sz}^{a}+F_{cz}^{a}\right)$$
$$-n_a\rho_a g-n_w\rho_w g+\rho g=0 \tag{5.49}$$

Compare Equation (5.44) and (5.45) to Equations (2.14) and (2.15) (i.e., the equilibrium equations for pore water and pore air in this book), respectively, and the following correlations can be established:

$$F_{sz}^{w}+F_{cz}^{w}=f_{swz},F_{sz}^{a}+F_{cz}^{a}=f_{saz},-n_a\rho_a g-n_w\rho_w g+\rho g=X_{sz} \tag{5.50}$$

Substituting Equation (5.50) into (5.49), and then:

$$\frac{\partial \tau_{xz}}{\partial x}+\frac{\partial \tau_{yz}}{\partial y}+\frac{\partial \sigma_{tz}}{\partial z}-n_a\frac{\partial u_a}{\partial z}-n_w\frac{\partial u_w}{\partial z}-f_{swz}-f_{saz}+X_{sz}=0 \tag{5.51}$$

Equation (5.51) is the equilibrium equation for the soil skeleton expressed by the total stress derived with the total stress equilibrium equation and the equilibrium equation for pore water, pore air and contractile skin provided by Fredlund.

In this book, the equilibrium equation for the soil skeleton expressed by effective stress is:

$$\frac{\partial \tau_{xz}}{\partial x}+\frac{\partial \tau_{yz}}{\partial y}+\frac{\partial \sigma_{z}}{\partial z}+\frac{1-n}{n}\frac{\partial (n_w u_w)}{\partial z}+\frac{1-n}{n}\frac{\partial (n_a u_a)}{\partial z}-f_{swz}-f_{saz}+X_{sz}=0 \tag{5.52}$$

which can be written as:

$$\frac{\partial \tau_{xz}}{\partial x}+\frac{\partial \tau_{yz}}{\partial y}+\frac{\partial \sigma_{z}}{\partial z}+\frac{1}{n}\left[\frac{\partial (n_w u_w)}{\partial z}+\frac{\partial (n_a u_a)}{\partial z}\right]-n_a\frac{\partial u_a}{\partial_z}-n_w\frac{\partial u_w}{\partial_z}$$
$$-f_{swz}-f_{saz}+X_{sz}=0 \tag{5.53}$$

It can be obtained by comparing Equation (5.51) and (5.53) that:

$$\frac{\partial \sigma_{tz}}{\partial z}=\frac{\partial \sigma_{z}}{\partial z}+\frac{1}{n}\left[\frac{\partial (n_w u_w)}{\partial z}+\frac{\partial (n_a u_a)}{\partial z}\right] \tag{5.54}$$

i.e.:

$$\frac{\partial \sigma_{tz}}{\partial z} = \frac{\partial}{\partial z}\left[\sigma_z + \frac{1}{n}\left(n_w u_w + n_a u_a\right)\right] \tag{5.55}$$

and then:

$$\sigma_{tz} = \sigma_z + \frac{1}{n}\left(n_w u_w + n_a u_a\right) \tag{5.56}$$

Since $S = \frac{n_w}{n}$, Equation (5.56) can be revised to:

$$\sigma_{tz} = \sigma_z - u_a + S(u_a - u_w) \tag{5.57}$$

Note that that Equation (5.57) is the effective stress equation for unsaturated soils. It is derived by comparing the total stress-expressed equilibrium equation for the soil skeleton given by Fredlund to the effective stress-expressed counterpart in this book.

Units and symbols

Table A List of symbols

Symbol	*Unit*	*Description*
A	m^2	Area
A_s	m^2	Area of the soil skeleton
A_{sa}	m^2	Soil skeleton area occupied by pore air
A_{sw}	m^2	Soil skeleton area occupied by pore water
A	–	Index of the force between pore water and the soil skeleton in the unit volume of the soil body
a_c	–	Ratio of contact area of skeleton particles to area of soil
$a_w = a/n_w$	–	Factor of soil-water interaction force for the unit volume of pore water
C	–	Volumetric compression coefficient of soil
C_s	–	Volumetric compression coefficient of soil particle
c'	kPa	Effective cohesion
D_{50}	mm	Diameter of 50% passing
E	kPa	Young modulus
F_c^a	kN/m^3	Interaction force between air and contractile skin
F_c^w	kN/m^3	Interaction force between water and contractile skin
F_s^a	kN/m^3	Interaction force between air and soil particles
F_s^w	kN/m^3	Interaction force between water and soil particles
f	kN/m^3	Force of soil-water interaction or seepage force
f_{sa}	kN/m^3	Seepage force of pore air on the soil skeleton
f_{sw}	kN/m^3	Seepage force or seepage force combined with unbalanced surface tension on the soil skeleton
f_{swa}	kN/m^3	Interaction force between the soil skeleton and pore water containing air bubbles
f_{ws}	kN/m^3	Seepage force or seepage force combined with unbalanced surface tension on pore water
f^*	–	Interaction function between the contractile skin and the soil structure
G_s	–	Specific gravity of solids
g	m/s^2	Gravity acceleration
H	m	Piezometer head or total head

(Continued)

Table A (Continued)

Symbol	*Unit*	*Description*
i	–	Hydraulic gradient or water potential gradient
J	kN/m^3	Seepage force
K_0	–	Coefficient of earth pressure at rest
k	m/s	Coefficient of permeability of the soil in saturated state
k_u	m/s	Coefficient of permeability of the soil in unsaturated state
n	%	Porosity
n_a	%	Porosity with respect to air phase
n_e	%	Effective porosity
n_{ew}	%	Porosity with respect to pore water excluding soil skeleton water
n_w	%	Porosity with respect to pore water
q_{max}	kPa	Maximum deviatoric stress
S	%	Degree of saturation
S_e	%	Effective degree of saturation
S_r	%	Residual degree of saturation
T_s	mN/m	Surface tension of water
t	s	Time
u^*	kPa	Equivalent pore fluid pressure
u_a	kPa	Pore air pressure
u_w	kPa	Pore water pressure
$u_a - u_w$	kPa	Matric suction
V	m^3	Volume
V_v	m^3	Volume of voids
v	m/s	Sectional average velocity
v'	m/s	Flow velocity at porous area
X_{sw}	kN/m^3	Body force of soil mass
X_{wa}	kN/m^3	Body force on the unit volume of the pore water containing air bubbles
γ	kN/m^3	Total unit weight
γ_{sat}	kN/m^3	Saturated unit weight
γ_d	kN/m^3	Dry unit weight
γ_w	kN/m^3	Unit weight of water
γ_a	kN/m^3	Unit weight of air
ε	%	Strain
ε_v	%	Volumetric strain
ν	–	Poisson's ratio
ρ	kg/m^3	Natural density
ρ_{sat}	kg/m^3	Saturated density
ρ_d	kg/m^3	Dry density
ρ_w	kg/m^3	Density of water
ρ_a	kg/m^3	Density of air

Symbol	*Unit*	*Description*
σ	kPa	Normal stress of the soil skeleton due to external forces
σ'	kPa	Effective stress
σ^s	kPa	Stress of the soil skeleton
σ_t	kPa	Total normal stress
τ	kPa	Shear stress
τ_f	kPa	Shear strength
φ'	degree	Effective angle of internal friction
χ	decimal	Bishop stress parameter
ψ	degree	Internal friction angle for the contact interface of soil particle

Subscript 0 in the equations in Section 4.2.1 indicates the volumetric water content distribution in the soil layers above the ground water level in static equilibrium condition.

References

Aitchison, G.D. (1965) Soils properties, shear strength, and consolidation. *Proceedings of the 6th International Conference on Soil Mechanics and Foundation Engineering*, University of Toronto Press, Toronto.

Aitchison, G.D. (1967) Separate roles of site investigation, quantification of soil properties, and selection of operational environment in the determination of foundation design on expensive soils. *Proceedings of the 3rd Asian Regional Conference on Soil Mechanics and Foundation Engineering*, Haifa, Israel. Vol. 3, pp. 72–77.

Aitchison, G.D. & Woodburn, J.A. (1969) Soil suction in foundation design. *Proceedings of the 7th International Conference on Soil Mechanics And Sound Engineering*, Mexico. Vol. 2, pp. 1–8.

Alonso, E.E., Gen, A. & Josa, A. (1990) The constitutive model for partially saturated soils. *Géotechnique*, 40(3), 405–430.

Alonso, E.E., Pereira, J.M., Vanunat, J. & Olovella, S. (2010) A microstructurally based effective stress for unsaturated soils. *Géotechnique*, 60(12), 913–925.

Barden, L., Madedor, A.O. & Sides, G.R. (1969) Volume change characteristics of unsaturated clay. *ASCE Journal of Soil Mechanics and Foundations Division*, 95, 33–52.

Biot, M.A. (1941) General theory of three-dimensional consolidation. *Journal of Applied Physics*, 12(2), 155–164.

Bishop, A.W. (1959) The principle of effective stress. *Teknisk Ukeblad*, 106(39), 113–143.

Bishop, A.W. & Blight, G.E. (1963) Some aspects of effective stress in saturated and partly saturated soils. *Géotechnique*, 13(3), 177–197.

Bolzon, G., Schrefler, A. & Zienkiewicz, O.C. (1996) Elasto-plastic soil constitutive laws generalized to partially saturated state. *Géotechnique*, 48(2), 279–289.

Brackley, I.J.A. (1971) Partial collapse in unsaturated expansive clay. *Proceedings of 5th Regional Conference on Soil Mechanics and Foundation Engineering*, South Africa.

Braja, M.D. (2008) *Advanced Soil Mechanics*, 3rd ed. Taylor & Francis, New York. p. 239.

Burland, J.B. (1965) Some aspects of the mechanical behaviour of partly saturated soils. *Moisture Equilibria and Moisture Change in Soils Beneath Covered Areas*, Butterworth, Sydney, Australia.

Carroll, M.M. (1980) Mechanical response of fluid-saturated porous materials. *Proceeding of the 15th International Congress of Theoretical and Applied Mechanics*, North-Holland, New York. pp. 251–261.

Chen, Z.H. & Qin, B. (2012) On stress state variables of unsaturated soils. *Rock and Soil Mechanics*, 33(1), 1–11. (in Chinese)

Chen, Z.Y., Zhou, J.X. & Wang, H.J. (1994) *Soil Mechanics*. Tsinghua University Press, Beijing. (in Chinese)

Coleman, G.E. (1967) Stress-strain relation for partly saturated soils. *Géotechnique*, 12(4), 348–350.

Croney, D. & Coleman, J.D. (1961) Pore pressure and suction in soil. *Pore Pressure & Suction in Soils Butterworths*, 31–37.

Croney, D., Coleman, J.D. & Black, W.P. (1958) *Movement and distribution of water in soil in relation to highway design and performance*. Highway Research Board Special Report.

Cui, Y.J. & Delage, P. (1996) Yielding and plastic behaviour of an unsaturated compacted silt. *Géotechnique*, 46(46), 291–311.

Curry, Q.C. (1982a) The essence of the hydration of clay mineral and hydrophilic of clay rocks. In: Li, S., *et al.* (eds) *Translation of Bonding Water in Soils*. Geological Press, Beijing. (in Chinese)

Curry, Q.C. (1982b) Temperature factors of physical-mechanical and physical-chemical properties of saturated clay with the density. In: Li, S., *et al.* (eds) *Translation of Bonding Water in Soils*. Geological Press, Beijing. (in Chinese)

Curry, Q.C. (1982c) Study on effects of osmosis phenomenon when the clay swelling. In: Li, S., *et al.* (eds) *Translation of Bonding Water in Soils*. Geological Press, Beijing. (in Chinese)

Curry, Q.C. (1982d) Influence of different types of bonding water on clay strength. In: Li, S., *et al.* (eds) *Translation of Bonding Water in Soils*. Geological Press, Beijing. (in Chinese)

Darcy, H. (1856) *Les Fontaines Publiques de la Ville de Dijon*. Victor Dalmont, Paris.

Feng, Y. (1984) *Introduction to Continuum Mechanics*. Science Press, Beijing. (in Chinese)

Fredlund, D.G. & Morgenstern, N.R. (1977) Stress state variables for unsaturated soils. *Journal of Geotechnical Engineering Division* (ASCE), 103(5), 447–466.

Fredlund, D.G. & Rahardjo, H. (1993) *Soil Mechanics for Unsaturated Soils*. John Wiley & Sons, Inc., Hoboken, NJ.

Fung, Y.C. (1977) *A First Course in Continuum Mechanics*, 2nd ed. Prentice-Hall, Englewood Cliffs, NJ. 340pp.

Gallipoli, D., Gens, A., Sharma, R. & Vaunat, J. (2003) An elasto-plastic model for unsaturated soil incorporating the effects of suction and degree of saturation on mechanical behavior. *Géotechnique*, 53(1), 123–135.

Herle, I. & Gudehus, G. (1999) Determination of parameters of a hypoplastic constitutive model from properties of grain assemblies. *Mechanics of Cohesive-Frictional Materials*, 4(5), 461–486.

Hossain, M.A. & Yin, J.H. (2010) Shear strength and dilative characteristics of an unsaturated compacted completely decomposed granite soil. *Canadian Geotechnical Journal*, 47, 1112–1126.

Houlsby, G.T. (1997) The work input to an unsaturated granular material. *Géotechnique*, 47(1), 193–196.

Jennings, J.E.B. (1960) A revised effective stress law for use in the prediction of the behavior of unsaturated soils. *Proceeding of the Conference on Pore Pressures*, Butterworths, London.

Jennings, J.E.B. & Burland, J.B. (1962) Limitations to the use of effective stresses in partly saturated soils. *Géotechnique*, 12(2), 125–144.

Kayadelen, C., Tekinsoy, M.A. & Taskiran, T. (2007) Influence of matric suction on shear behavior of a residual clayey soil. *Environmental Geology*, 53, 891–901.

Khalili, N., Geiser, F. & Blight, G.E. (2004) Effective stress in unsaturated soils: review with new evidence. *International Journal of Geomechanics*, 4, 115–126.

Khalili, N. & Khabbaz, M.H. (1998) A unique relationship for χ for the determination of the shear strength of unsaturated soils. *Géotechnique*, 48(5), 681–687.

Kohgo, Y., Nakano, M. & Miyazaki, T. (1993) Theoretical aspects of constitutive modelling for unsaturated soils. *Soils and Foundations*, 33(4), 49–63.

Lade, P.V. & Boer, R.D.E. (1997) The concept of effective stress for soil concrete and rock. *Géotechnique*, 47(1), 61–78.

Lee, I.M., Sung, S.G. & Cho, G.C. (2005) Effect of stress state on the unsaturated shear strength of a weathered granite. *Canadian Geotechnical Journal*, 42(2), 624–631.

Lei, Z.D., Yang, S.X. & Xie, S.C. (1988) *Soil Water Dynamics*. Tsinghua University Press, Beijing. (in Chinese)

Li, G. (2011) Some problems about principle of effective stress. *Chinese Journal of Geotechnical Engineering*, 33(2), 315–320. (in Chinese)

Li, X.S. (2007) Thermodynamics-based constitutive framework for unsaturated soils. *Gèotechnique*, 57(5), 411–422.

Loret, B. & Khalili, N. (2000) A three phase model for unsaturated soils. *International Journal of Numerical and Analytical Methods in Geomechanics*, 24, 893–927.

Loret, B. & Khalili, N. (2002) An effective stress elastic-plastic model for unsaturated porous media. *Mechanics of Materials*, 34(2), 97–116.

Lu, N. (2008) Is matric suction stress variable? *Journal of Geotechnical and Geoenvironmental Engineering* (ASCE), 134(7), 899–905.

Lu, N., Godt, J.W. & Wu, D.T. (2010) A closed-form equation for effective stress in unsaturated soil. *Water Resources Research*, 46, 1–14.

Matyas, E.L. & Radhakrishna, H.S. (1968) Volumetric change characteristics of partially saturated soils. *Gèotechnique*, 18(4), 432–448.

Miao, L.C., Cui, Y.J. & Cui, Y. (2015) Hydromechanical behaviors of unsaturated soils. *Journal of Materials in Civil Engineering*, 27(7), 1–9.

Mitchell, J.K. (1993) *Fundamentals of Soil Behavior*. John Wiley & Sons Inc., New York.

Modaressi, A. & Abou-Bekr, N. (1994) A unified approach to model the behavior of saturated and unsaturated soils. In: Siriwardane, H.J. & Zaman, M.M. (eds) *Computer Methods and Advances in Geomechanics, Proceedings of the Eighth International Conference on Computer Methods and Advances in Geomechanics*, Balkema, Amsterdam. pp. 1507–1513.

Morgenstern, N.R. (1979) Properties of compacted soils. *Proceedings of the 6th Pan-American Conference on Soil Mechanics and Foundation Engineering*. Vol. 3, pp. 349–354.

Ng, C.W.W. & Chiu, C.F. (2003) A state-dependent elasto-plastic model for saturated and unsaturated soils. *Gèotechnique*, 53(9), 809–829.

Nur, A. & Byerlee, J.D. (1971) An exact effective stress law for elastic deformation of rock with fluids. *Journal of Geophysical Research*, 76, 6414–6419.

Olivier, C. (2004) *Poro Mechanics*. John Wiley & Sons, Ltd., Chichester, UK.

Reint, D.B. (2005) *Trends in Continuum Mechanics of Porous*. Springer, Dordrecht, The Netherlands.

Richards, B.G. (1966) The significance of moisture flow and equilibria in unsaturated soils in relation to the design of engineering structures built on shallow foundations in Australia. Presented at the *Symposium on Permeability and Capillary*, American Society for Testing and Materials, Atlantic City, NJ.

Shao, L.T. (1996) *Fundamental Theory of Porous Material Mechanics and Its Application in Soil Mechanics*. Doctoral Dissertation, Dalian University of Technology, Dalian, China. (in Chinese)

Shao, L.T. (2000) *Finite element stability analysis of soil and some problems in the research of the theory of unsaturated soil mechanics*. Postdoctoral Research Report, Dalian University of Technology, Dalian, China. (in Chinese)

Shao, L.T. (2011a) *Research and Exploration of Soil Mechanics*. Science Press, Beijing. (in Chinese)

Shao, L.T. (2011b) Skeleton stress equation for saturated soils. *Chinese Journal of Geotechnical Engineering*, 33(12), 1833–1837. (in Chinese)

Shao, L.T. (2012) *Research and Exploration of Soil Mechanics-5 Lectures in the New System of Soil Mechanics Theory*. Science Press, Beijing. (in Chinese)

Shao, L.T. & Guo, X.X. (2014) *New Explanation of Effective Stress*. China Water & Power Press, Beijing. (in Chinese)

Shao, L.T., Zheng, G.F., Guo, X.X. & Liu, G. (2014) Principle of effective stress for unsaturated soils. *Proceeding of the 6th International Conference on Unsaturated Soils*, Sydney, Australia. pp. 239–245.

Shen, Z. (1995) Discussion on consolidation theory and effective stress. *Chinese Journal of Geotechnical Engineering*, 17(6), 118–119. (in Chinese)

Sheng, D.C., Smith, D.W., Sloan, S.W. & Gens, A. (2003) Finite element formulation and algorithms for unsaturated soils, Part I: Theory. *International Journal for Numerical & Analytical Methods in Geomechanics*, 27(9), 745–765.

Sheng, D.C., Zhou, A.N. & Fredlund, D.G. (2011) Shear strength criteria for unsaturated soils. *Geotechnical and Geological Engineering*, 29, 145–159.

Skempton, A.W. (1961) Effective stress in soils, concrete and rocks, pore pressure and suction in soils. *Conference organized by the British National Society of International Society of Soil Mechanics and Foundation Engineering*, Butterworths, London. pp. 4–25.

Terzaghi, K. (1925) *Soil Mechanics on Soil Physics Basis*. Deuticke, Vienna.

Terzaghi, K. (1936) The shearing resistance of saturated soils and the angle between the planes of shear. *Proceedings for the 1st International Conference on Soil Mechanics and Foundation Engineering*, Cambridge, MA. Vol. 1, pp. 54–56.

Vanapalli, S.K., Fredlund, D.G., Pufahl, D.E. & Clifton, A.W. (1996) Model for the prediction of shear strength with respect to soil suction. *Canadian Geotechnical Journal*, 33(3), 379–392.

Wheeler, S.J. & Sivakumar, V. (1995) An elasto-plastic critical state framework for unsaturated soil. *Géotechnique*, 45(1), 35–53.

Wheeler, S.J. & Sivakumar, V. (2003) Influence of compaction procedure on the mechanical behaviour of an unsaturated compacted clay. *Géotechnique*, 53(9), 842–843.

Yin, Z.Z., *et al.* (2007) *Principles of Geotechnical Engineering*. China Water & Power Press, Beijing. (in Chinese)

Zhan, L.T. & Ng, C.W. (2006) Shear strength characteristics of an unsaturated expansive clay. *Canadian Geotechnical Journal*, 43, 751–763.

Zhao, C. & Cai, G. (2009) Principles of generalized effective stress for unsaturated soils. *Rock and Soil Mechanics*, 30(11), 3233–3236. (in Chinese)

Zhao, C., Li, J. & Liu, Y. (2013) Discussion on some fundamental problems in unsaturated soil mechanics. *Rock and Soil Mechanics*, 34(7), 1825–1831. (in Chinese)

Zienkiewicz, O.C. & Shiomi, T. (1984) Dynamic behavior of saturated porous media: the generalized Boit formulation and its numerical solution. *International Journal for Numerical and Analytical Methods in Geomechanics*, 8(1), 71–96.

Subject Index

Printed and bound by CPI Group (UK) Ltd, Croydon, CR0 4YY
24/10/2024
01778286-0013